Human Biology

Human Biology

Gary Hansen

R CALLISTO REFERENCE

www.callistoreference.com

Callisto Reference,
118-35 Queens Blvd., Suite 400,
Forest Hills, NY 11375, USA

Visit us on the World Wide Web at:
www.callistoreference.com

ISBN: 978-1-64116-235-7 (Hardback)

Cataloging-in-Publication Data

Human biology / Gary Hansen.
    p. cm.
Includes bibliographical references and index.
ISBN 978-1-64116-235-7
1. Human biology. 2. Biology. 3. Physical anthropology. I. Hansen, Gary.
QP34.5 .H86 2019
612--dc23

Table of Contents

Preface

Human biology is the science that studies the human body through the lens of anatomy, human genetics and evolution, physiology, immunology, epidemiology and anthropology. It also provides the foundation for a scientific approach to the study of diseases in humans, which involves therapy, diagnosis and prevention. The important sub-disciplines of human biology are pathophysiology, medical genetics, pharmacology, toxicology, pathology, etc. Some of the principal systems of the human body are the circulatory system, digestive system, the nervous system, respiratory system, muscular system and the skeletal system. This textbook attempts to understand the science of human biology in an interdisciplinary manner. This book is a valuable compilation of topics, ranging from the basic to the most complex theories and principles in the field of human biology and related fields. Coherent flow of topics, student-friendly language and extensive use of examples make this book an invaluable source of knowledge.

A foreword of all Chapters of the book is provided below:

Chapter 1 - Human anatomy is the science concerned with the study of the shape and form of the human body. It is made of a number of organs and organ systems. This is an introductory chapter, which will introduce briefly al the significant aspects of the human anatomy and organs; **Chapter 2 -** The smooth functioning of the human body is achieved through different kinds of cells, tissues and organ systems. The primary organ systems fundamental to the working of the human body are the cardiovascular system, nervous system, digestive system, immune system, respiratory system, etc. which have been discussed in elaborate detail in this chapter; **Chapter 3 -** Human genetics delves into the study of the inheritable traits which have been passed on through generations. Evolution and reproduction are two significant aspects of this field, which aid the understanding of human biology. The fundamental topics related to the understanding of human evolution, genetics and reproduction have been covered in this chapter, such as anatomy of bipedalism, behavioral modernity, evolution of human intelligence, human DNA, RNA, etc.; **Chapter 4 -** The human skeleton is the system that provides the framework to the human body and consists of around 206 bones in adulthood. The human skeleton can be divided into two groups, the axial skeleton and the appendicular skeleton, which have elaborately covered in this chapter. It further elucidates the constituting structures of the skeletal systems, such as shoulder girdle, pelvis, thighs, ankles, human head, etc.

I would like to thank the entire editorial team who made sincere efforts for this book and my family who supported me in my efforts of working on this book. I take this opportunity to thank all those who have been a guiding force throughout my life.

Gary Hansen

Chapter 1

Understanding Human Anatomy

Human anatomy is the science concerned with the study of the shape and form of the human body. It is made of a number of organs and organ systems. This is an introductory chapter, which will introduce briefly all the significant aspects of the human anatomy and organs.

Anatomy is the identification and description of the structures of living things. It is a branch of biology and medicine.

The study of anatomy goes back over 2,000 years, to the Ancient Greeks. It can be divided into three broad areas: Human anatomy, zootomy, or animal anatomy, and phytotomy, which is plant anatomy.

Human anatomy is the study of the structures of the human body. An understanding of anatomy is key to the practice of health and medicine.

The word "anatomy" comes from the Greek words "ana," meaning "up," and "tome," meaning "a cutting." Studies of anatomy have traditionally depended on cutting up, or dissection, but now, with imaging technology, it is increasingly possible to see how a body is made up without dissection.

In its broadest sense, anatomy is the study of the structure of an object, in this case the human body. Human anatomy deals with the way the parts of humans, from molecules to bones, interact to form a functional unit. The study of anatomy is distinct from the study of physiology, although the two are often paired. While anatomy deals with the structure of an organism, physiology deals with the way the parts function together. For example, an anatomist may study the types of cells in the cardiac conduction system and how those cells are connected, while a physiologist would look at why and how the heart beats. Thus, anatomy and physiology are separate, but complimentary, studies of how an organism works.

As in veterinary anatomy, human anatomy is subdivided into macroscopic (or gross) and microscopic anatomy.

- Macroscopic anatomy describes structures, organs, muscles, bones etc. which are visible to the naked eye, that is macroscopic. In order to establish a certain order they are divided topographically and systematically.

- Microscopic human anatomy is the „study of tissues", that is histology. It may be further separated into cytology, the pure study of cells. In contrast to macroscopic anatomy you require – as the name suggests - an optical magnification in order to evaluate microscopic (e.g. cellular) structures.

- Embryology needs to be considered as part of human anatomy as well. It is the study of the development of the human body beginning from fertilization of the ovum until birth.

Vital Organs

Humans have five vital organs that are essential for survival. These are the brain, heart, kidneys, liver and lungs.

The human brain is the body's control center, receiving and sending signals to other organs through the nervous system and through secreted hormones. It is responsible for our thoughts, feelings, memory storage and general perception of the world.

The human heart is a responsible for pumping blood throughout our body.

The job of the kidneys is to remove waste and extra fluid from the blood. The kidneys take urea out of the blood and combine it with water and other substances to make urine.

The liver has many functions, including detoxifying of harmful chemicals, breakdown of drugs, filtering of blood, secretion of bile and production of blood-clotting proteins.

The lungs are responsible for removing oxygen from the air we breathe and transferring it to our blood where it can be sent to our cells. The lungs also remove carbon dioxide, which we exhale.

Human Body

Human body, the physical substance of the human organism, composed of living cells and extra-cellular materials and organized into tissues, organs, and systems.

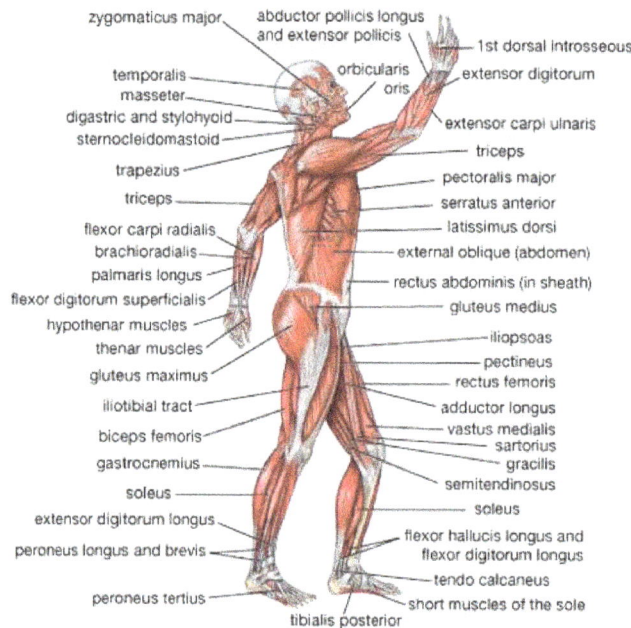

Lateral view of the human muscular system

Humans are, of course, animals—more particularly, members of the order Primates in the subphylum Vertebrata of the phylum Chordata. Like all chordates, the human animal has a bilaterally symmetrical body that is characterized at some point during its development by a dorsal supporting rod (the notochord), gill slits in the region of the pharynx, and a hollow dorsal nerve cord. Of these features, the first two are present only during the embryonic stage in the human; the notochord is replaced by the vertebral column, and the pharyngeal gill slits are lost completely. The dorsal nerve cord is the spinal cord in humans; it remains throughout life.

Characteristic of the vertebrate form, the human body has an internal skeleton that includes a backbone of vertebrae. Typical of mammalian structure, the human body shows such characteristics as hair, mammary glands, and highly developed sense organs.

Beyond these similarities, however, lie some profound differences. Among the mammals, only humans have a predominantly two-legged (bipedal) posture, a fact that has greatly modified the general mammalian body plan. (Even the kangaroo, which hops on two legs when moving rapidly, walks on four legs and uses its tail as a "third leg" when standing.) Moreover, the human brain, particularly the neocortex, is far and away the most highly developed in the animal kingdom. As intelligent as are many other mammals—such as chimpanzees and dolphins—none have achieved the intellectual status of the human species.

Composition

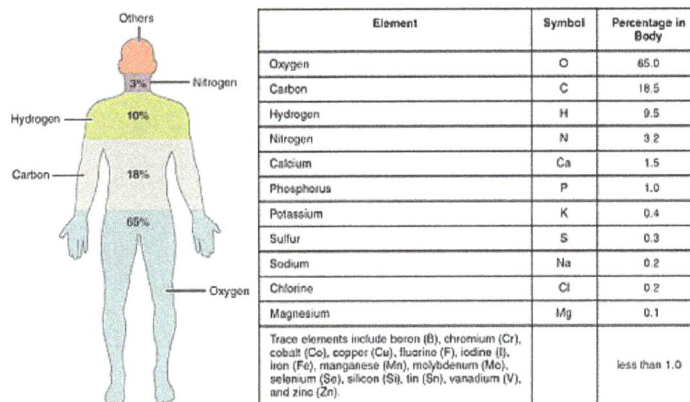

Element	Symbol	Percentage in Body
Oxygen	O	65.0
Carbon	C	18.5
Hydrogen	H	9.5
Nitrogen	N	3.2
Calcium	Ca	1.5
Phosphorus	P	1.0
Potassium	K	0.4
Sulfur	S	0.3
Sodium	Na	0.2
Chlorine	Cl	0.2
Magnesium	Mg	0.1
Trace elements include boron (B), chromium (Cr), cobalt (Co), copper (Cu), fluorine (F), iodine (I), iron (Fe), manganese (Mn), molybdenum (Mo), selenium (Se), silicon (Si), tin (Sn), vanadium (V), and zinc (Zn).		less than 1.0

Elements of the human body by mass. Trace elements are less than 1% combined (and each less than 0.1%)

The human body is composed of elements including hydrogen, oxygen, carbon, calcium and phosphorus. These elements reside in trillions of cells and non-cellular components of the body.

The adult male body is about 60% water for a total water content of some 42 litres. This is made up of about 19 litres of extracellular fluid including about 3.2 litres of blood plasma and about 8.4 litres of interstitial fluid, and about 23 litres of fluid inside cells. The content, acidity and composition of the water inside and outside cells is carefully maintained. The main electrolytes in body water outside cells are sodium and chloride, whereas within cells it is potassium and other phosphates.

Cells

The body contains trillions of cells, the fundamental unit of life. At maturity, there are roughly 30–37 trillion cells in the body, an estimate arrived at by totalling the cell numbers of all the organs of the body and cell types. The body is also host to about the same number of non-human cells as well as multicellular organisms which reside in the gastrointestinal tract and on the skin. Not all parts of the body are made from cells. Cells sit in an extracellular matrix that consists of proteins such as collagen, surrounded by extracellular fluids. Of the 70 kg weight of an average human body, nearly 25 kg is non-human cells or non-cellular material such as bone and connective tissue.

Cells in the body function because of DNA. DNA sits within the nucleus of a cell. Here, parts of DNA are copied and sent to the body of the cell via RNA. The RNA is then used to create proteins which form the basis for cells, their activity, and their products. Proteins dictate cell function and gene expression, a cell is able to self-regulate by the amount of proteins produced. However, not all cells have DNA – some cells such as mature red blood cells lose their nucleus as they mature.

Tissues

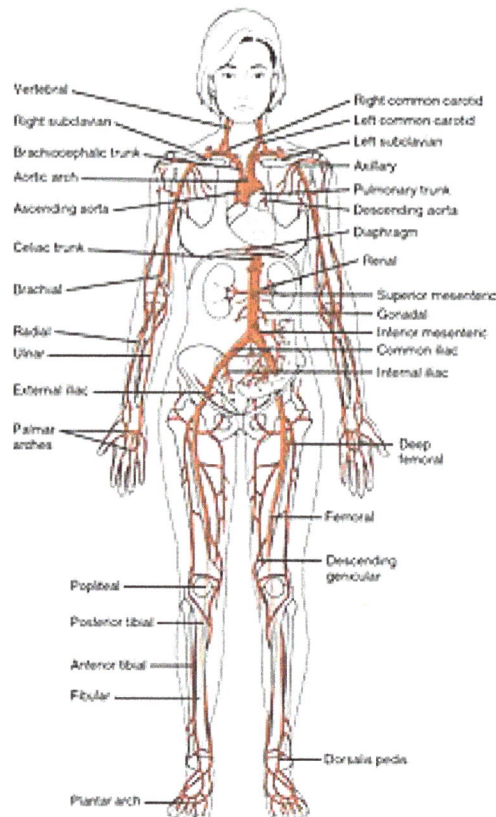

Human Body

The body consists of many different types of tissue, defined as cells that act with a specialised function. The study of tissues is called histology and often occurs with a microscope. The body consists of four main types of tissues – lining cells (epithelia), connective tissue, nervous tissue and muscle tissue.

Cells that lie on surfaces exposed to the outside world or gastrointestinal tract (epithelia) or internal cavities (endothelium) come in numerous shapes and forms – from single layers of flat cells, to cells with small beating hair-like cilia in the lungs, to column-like cells that line the stomach. Endothelial cells are cells that line internal cavities including blood vessels and glands. Lining cells regulate what can and can't pass through them, protect internal structures, and function as sensory surfaces.

Organs

Organs, structured collections of cells with a specific function, sit within the body. Examples include the heart, lungs and liver. Many organs reside within cavities within the body. These cavities include the abdomen and pleura.

Systems

Circulatory System

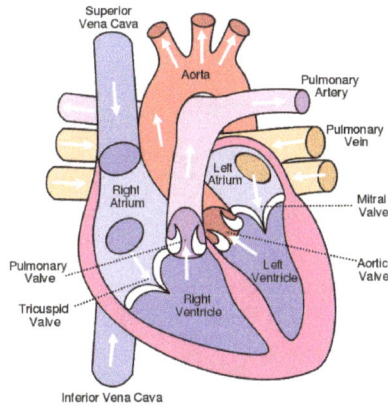

The circulatory system comprises the heart and blood vessels (arteries, veins and capillaries). The heart propels the circulation of the blood, which serves as a "transportation system" to transfer oxygen, fuel, nutrients, waste products, immune cells and signalling molecules (i.e., hormones) from one part of the body to another. The blood consists of fluid that carries cells in the circulation, including some that move from tissue to blood vessels and back, as well as the spleen and bone marrow.

Digestive System

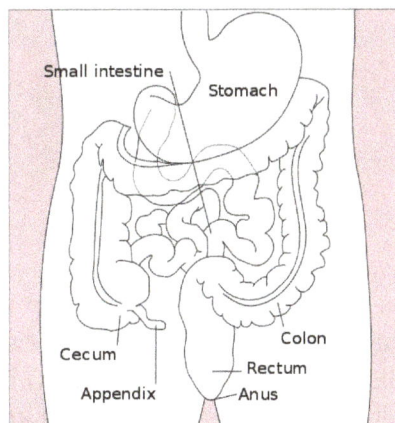

The digestive system consists of the mouth including the tongue and teeth, esophagus, stomach, (gastrointestinal tract, small and large intestines, and rectum), as well as the liver, pancreas, gallbladder, and salivary glands. It converts food into small, nutritional, non-toxic molecules for distribution and absorption into the body.

Endocrine System

The endocrine system consists of the principal endocrine glands: the pituitary, thyroid, adrenals, pancreas, parathyroids, and gonads, but nearly all organs and tissues produce specific endocrine hormones as well. The endocrine hormones serve as signals from one body system to another regarding an enormous array of conditions, and resulting in variety of changes of function.

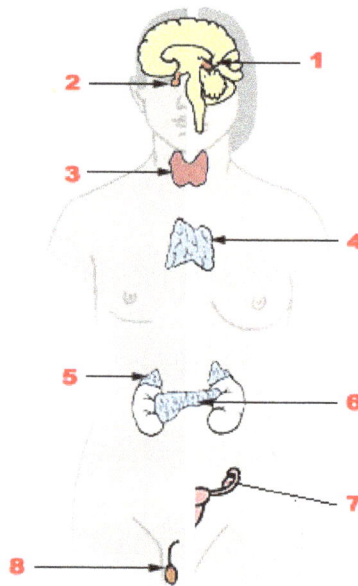

Immune System

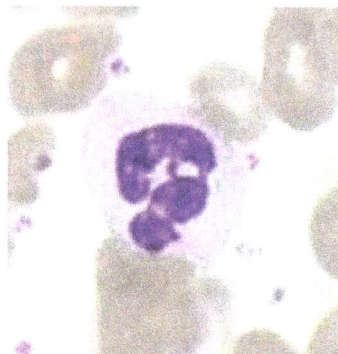

The immune system consists of the white blood cells, the thymus, lymph nodes and lymph channels, which are also part of the lymphatic system. The immune system provides a mechanism for the body to distinguish its own cells and tissues from outside cells and substances and to neutralize or destroy the latter by using specialized proteins such as antibodies, cytokines, and toll-like receptors, among many others.

Integumentary System

The integumentary system consists of the covering of the body (the skin), including hair and nails as well as other functionally important structures such as the sweat glands and sebaceous glands. The skin provides containment, structure, and protection for other organs, and serves as a major sensory interface with the outside world.

Lymphatic System

The lymphatic system extracts, transports and metabolizes lymph, the fluid found in between cells. The lymphatic system is similar to the circulatory system in terms of both its structure and its most basic function, to carry a body fluid.

Musculoskeletal System

The musculoskeletal system consists of the human skeleton (which includes bones, ligaments, tendons, and cartilage) and attached muscles. It gives the body basic structure and the ability for movement. In addition to their structural role, the larger bones in the body contain bone marrow, the site of production of blood cells. Also, all bones are major storage sites for calcium and phosphate. This system can be split up into the muscular system and the skeletal system.

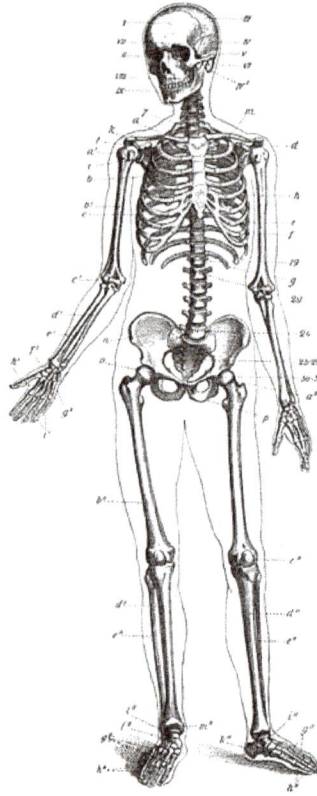

Nervous System

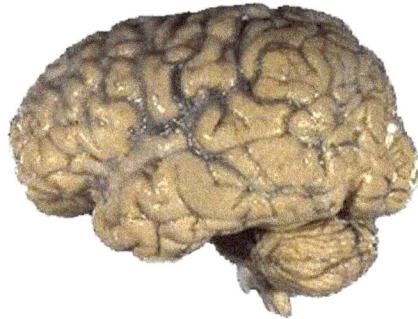

The nervous system consists of the central nervous system (the brain and spinal cord) and the peripheral nervous system consists of the nerves and ganglia outside the brain and spinal cord. The brain is the organ of thought, emotion, memory, and sensory processing, and serves many aspects of communication and controls various systems and functions. The special senses consist of vision, hearing, taste, and smell. The eyes, ears, tongue, and nose gather information about the body's environment.

Reproductive System

The reproductive system consists of the gonads and the internal and external sex organs. The reproductive system produces gametes in each sex, a mechanism for their combination, and in the female a nurturing environment for the first 9 months of development of the infant.

Respiratory System

The respiratory system consists of the nose, nasopharynx, trachea, and lungs. It brings oxygen from the air and excretes carbon dioxide and water back into the air.

Urinary System

The urinary system consists of the kidneys, ureters, bladder, and urethra. It removes toxic materials from the blood to produce urine, which carries a variety of waste molecules and excess ions and water out of the body.

Anatomy

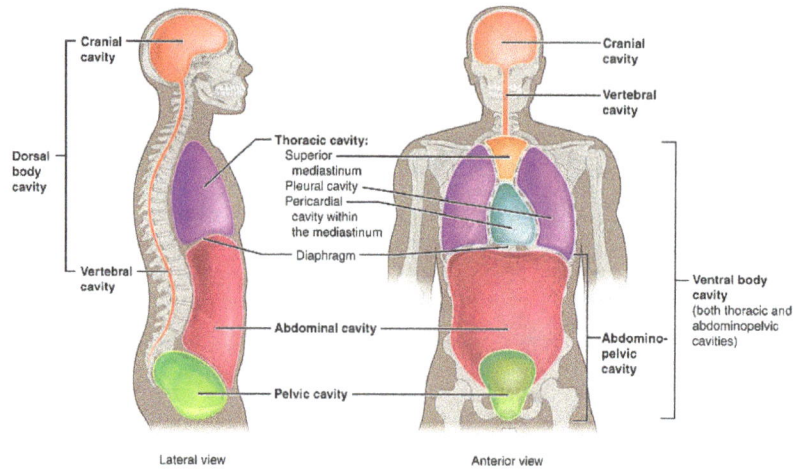

Cavities of human body

Human anatomy is the study of the shape and form of the human body. The human body has four limbs (two arms and two legs), a head and a neck which connect to the torso. The body's shape is determined by a strong skeleton made of bone and cartilage, surrounded by fat, muscle, connective tissue, organs, and other structures. The spine at the back of the skeleton contains the flexible vertebral column which surrounds the spinal cord, which is a collection of nerve fibres connecting the brain to the rest of the body. Nerves connect the spinal cord and brain to the rest of the body. All major bones, muscles, and nerves in the body are named, with the exception of anatomical variations such as sesamoid bones and accessory muscles.

Blood vessels carry blood throughout the body, which moves because of the beating of the heart. Venules and veins collect blood low in oxygen from tissues throughout the body. These collect in progressively larger veins until they reach the body's two largest veins, the superior and inferior vena cava, which drain blood into the right side of the heart. From here, the blood is pumped into the lungs where it receives oxygen and drains back into the left side of the heart. From here, it is pumped into the body's largest artery, the aorta, and then progressively smaller arteries and arterioles until it reaches tissue. Here blood passes from small arteries into capillaries, then small veins and the process begins again. Blood carries oxygen, waste products, and hormones from one place in the body to another. Blood is filtered at the kidneys and liver.

The body consists of a number of different cavities, separated areas which house different organ systems. The brain and central nervous system reside in an area protected from the rest of the body by the blood brain barrier. The lungs sit in the pleural cavity. The intestines, liver, and spleen sit in the abdominal cavity

Height, weight, shape and other body proportions vary individually and with age and sex. Body shape is influenced by the distribution of muscle and fat tissue.

Physiology

Human physiology is the study of how the human body functions. This includes the mechanical, physical, bioelectrical, and biochemical functions of humans in good health, from organs to the cells of which they are composed. The human body consists of many interacting systems of organs. These interact to maintain homeostasis, keeping the body in a stable state with safe levels of substances such as sugar and oxygen in the blood.

Each system contributes to homeostasis, of itself, other systems, and the entire body. Some combined systems are referred to by joint names. For example, the nervous system and the endocrine system operate together as the neuroendocrine system. The nervous system receives information from the body, and transmits this to the brain via nerve impulses and neurotransmitters. At the same time, the endocrine system releases hormones, such as to help regulate blood pressure and volume. Together, these systems regulate the internal environment of the body, maintaining blood flow, posture, energy supply, temperature, and acid balance (pH).

Human development, the process of growth and change that takes place between birth and maturity.

Human growth is far from being a simple and uniform process of becoming taller or larger. As a child gets bigger, there are changes in shape and in tissue composition and distribution. In the newborn infant the head represents about a quarter of the total length; in the adult it represents about one-seventh. In the newborn infant the muscles constitute a much smaller percentage of the total body mass than in the young adult. In most tissues, growth consists both of the formation of new cells and the packing in of more protein or other material into cells already present; early in development cell division predominates and later cell filling.

Types and Rates of Human Growth

Different tissues and different regions of the body mature at different rates, and the growth and development of a child consists of a highly complex series of changes. It is like the weaving of a cloth whose pattern never repeats itself. The underlying threads, each coming off its reel at its own rhythm, interact with one another continuously, in a manner always highly regulated and controlled. The fundamental questions of growth relate to these processes of regulation, to the program that controls the loom, a subject as yet little understood. Meanwhile, height is in most circumstances the best single index of growth, being a measure of a single tissue (that of the skeleton; weight is a mixture of all tissues, and this makes it a less useful parameter in a long-term following of a child's growth). In this section, the height curves of girls and boys are considered in the three chief phases of growth; that is (briefly) from conception to birth, from birth until puberty, and during puberty. Also described are the ways in which other organs and tissues, such as fat, lymphoid tissue, and the brain, differ from height in their growth curves. There is a brief discussion of some of the problems that beset the investigator in gathering and analyzing data about growth of children, of the genetic and environmental factors that affect rate of growth and final size, and of the way hormones act at the various phases of the growth process. Lastly, there is a brief look at disorders of growth. Throughout, the emphasis is on ways in which individuals differ in their rates of growth and development.

The changes in height of the developing child can be thought of in two different ways: the height attained at successive ages and the increments in height from one age to the next, expressed as rate of growth per year. If growth is thought of as a form of motion, the height attained at successive ages can be considered the distance travelled, and the rate of growth, the velocity. The velocity or rate of growth reflects the child's state at any particular time better than does the height attained, which depends largely on how much the child has grown in all preceding years. The blood and tissue concentrations of those substances whose amounts change with age are thus more likely to run parallel to the velocity rather than to the distance curve. In some circumstances, indeed, it is the acceleration rather than the velocity curve that best reflects physiological events.

In general, the velocity of growth decreases from birth onward (and actually from as early as the fourth month of fetal life), but this decrease is interrupted shortly before the end of the growth period. At this time, in boys from about 13 to 15 years, there is marked acceleration of growth, called the adolescent growth spurt. From birth until age four or five, the rate of growth in height declines rapidly, and then the decline, or deceleration, gets gradually less, so that in some children the velocity is practically constant from five or six up to the beginning of the adolescent spurt. A slight increase in velocity is sometimes said to occur between about six and eight years.

This general velocity curve of growth in height begins a considerable time before birth. The peak velocity of length is reached at about four months after the mother's last menstruation. (Age in the fetal period is usually reckoned from the first day of the last menstrual period, an average of two weeks before actual fertilization, but, as a rule, the only locatable landmark.)

Growth in weight of the fetus follows the same general pattern as growth in length, except that the peak velocity is reached much later, at approximately 34 weeks after the mother's last menstrual period.

There is considerable evidence that from about 34 to 36 weeks onward the rate of growth of the fetus slows down because of the influence of the maternal uterus, whose available space is by then becoming fully occupied. Twins slow down earlier, when their combined weight is approximately the 36-week weight of a single fetus. Babies who are held back in this way grow rapidly as soon as they have emerged from the uterus. Thus there is a significant negative association between weight of a baby at birth and weight increment during the first year; in general, larger babies grow less, the smaller more. For the same reason there is practically no relation between adult size and the size of that person at birth, but a considerable relation has developed by the time the person is two years old. This slowing-down mechanism enables a genetically large child developing in the uterus of a small mother to be delivered successfully. It operates in many species of animals; the most dramatic demonstration was by crossing reciprocally a large Shire horse and a small Shetland pony. The pair in which the mother was a Shire had a large newborn foal, and the pair in which the mother was Shetland had a small foal. But both foals were the same size after a few months, and when fully grown both were about halfway between their parents. The same has been shown in cattle crosses.

Poor environmental circumstances, especially of nutrition, result in lowered birth weight in the human being. This seems chiefly to be caused by a reduced rate of growth in the last two to four weeks of fetal life, for weights of babies born in 36 or 38 weeks in various parts of the world in various circumstances are said to be similar. Mothers who, because of adverse circumstances in their own childhood, have not achieved their full growth potential may produce smaller fetuses than they would have, had they grown up in better circumstances. Thus two generations or even more may be needed to undo the effect of poor environmental circumstances on birth weight.

The great rate of growth of the fetus compared with that of the child is largely due to the fact that cells are still multiplying. The proportion of cells undergoing mitosis (the ordinary process of cell multiplication by splitting) in any tissue becomes progressively less as the fetus gets older, and it is generally thought that few if any new nerve cells (apart from the cells in the supporting tissue, or neuroglia) and only a limited proportion of new muscle cells appear after six postmenstrual months, the time when the velocity in linear dimensions is dropping sharply.

The muscle and nerve cells of the fetus are considerably different in appearance from those of the child or adult. Both have little cytoplasm (cell substance) around the nucleus. In the muscle there is a great amount of intercellular substance and a much higher proportion of water than in mature muscle. The later fetal and the postnatal growth of the muscle consists chiefly of building up the cytoplasm of the muscle cells; salts are incorporated and the contractile proteins formed. The cells become bigger, the intercellular substance largely disappears, and the concentration of water decreases. This process continues quite actively up to about three years of age and slowly thereafter; at adolescence it briefly speeds up again, particularly in boys, under the influence of androgenic (male sex) hormones. In the nerve cells cytoplasm is added and elaborated, and extensions grow that carry impulses from and to the cells—the axons and dendrites, respectively. Thus postnatal growth, for at least some tissues, is chiefly a period of development and enlargement of existing cells, while early fetal life is a period of division and addition of new cells.

Organ Anatomy

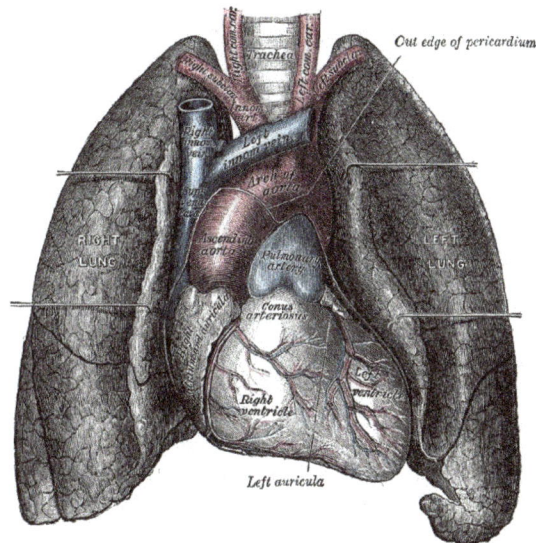

The heart and lungs

In biology, an organ (Latin: *organum,* "instrument, tool") is a group of tissues that perform a specific function or group of functions.

Organs, exemplified by such diverse components as brain, eyes, and liver, are one of several levels of organization in living organisms. A given organ is usually thought of as being a component of an organ system, a group of organs that work together to perform a set of related functions, such as the digestive system composed of the mouth, esophagus, intestines, and other organs. At a lower level of organization an organ is an aggregation of several tissues that interact to perform a specific function, such as the heart pumping blood or the stomach digesting food. In the case of the stomach, muscle tissue causes movement, epithelial tissue secretes enzymes, such as pepsin, and nervous tissue conducts messages back and forth between the stomach and brain (Towle 1989). A tissue, in turn, is an aggregation of interconnected, morphologically, and functionally similar cells, and associated intercellular matter, that together perform one or more specific functions within an organism.

In a living organism, cells depend upon other cells, tissues depend upon other tissues, and organs depend upon other organs to perform their particular functions to keep the entire organism alive. Each cell, tissue, and organ demonstrates the principle of bi-level functionality: The entity not only performs actions for its own maintenance, self-preservation, and self-strengthening, but also performs specific actions that contribute to the larger entity (the tissue, organ, organ system, or body). The body, on the other hand, supports the individual cell, tissue, organ, and organ system by providing access to food, oxygen, and shelter, and by disposing of waste materials.

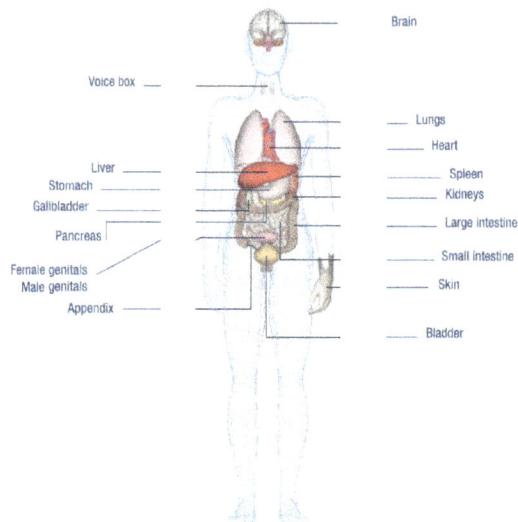

More human anatomy diagrams: nervous system, skeleton,
front view of muscles, back view of muscles Organise the
organs in our interactive body

Structure

Tissue

In biology, tissue is a cellular organizational level between cells and complete organs. A tissue is an ensemble of similar cells and their extracellular matrix from the same origin that together carry out a specific function. Organs are then formed by the functional grouping together of multiple tissues.

The study of human and animal tissues is known as histology or, in connection with disease, histo-pathology. For plants, the discipline is called plant anatomy. The classical tools for studying tissues are the paraffin block in which tissue is embedded and then sectioned, the histological stain, and the optical microscope. In the last couple of decades, developments in electron microscopy, immunofluorescence, and the use of frozen tissue sections have enhanced the detail that can be observed in tissues. With these tools, the classical appearances of tissues can be examined in health and disease, enabling considerable refinement of medical diagnosis and prognosis.

Organ Systems

Two or more organs working together in the execution of a specific body function form an organ system, also called a biological system or body system. The functions of organ systems often share significant overlap. For instance, the nervous and endocrine system both operate via a shared organ, the hypothalamus. For this reason, the two systems are combined and studied as the neuroendocrine system. The same is true for the musculoskeletal system because of the relationship between the muscular and skeletal systems.

Function

Animals

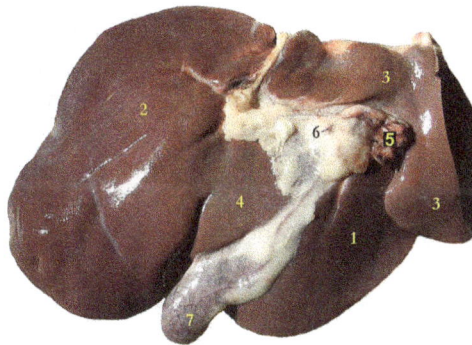

The liver and gallbladder of a sheep

Animals such as humans have a variety of organ systems. These specific systems are also widely studied in human anatomy.

- Cardiovascular system: pumping and channeling blood to and from the body and lungs with heart, blood and blood vessels.

- Digestive system: digestion and processing food with salivary glands, esophagus, stomach, liver, gallbladder, pancreas, intestines, colon, rectum and anus.

- Endocrine system: communication within the body using hormones made by endocrine glands such as the hypothalamus, pituitary gland, pineal body or pineal gland, thyroid, parathyroids and adrenals, i.e., adrenal glands.

- Excretory system: kidneys, ureters, bladder and urethra involved in fluid balance, electrolyte balance and excretion of urine.

- Lymphatic system: structures involved in the transfer of lymph between tissues and the blood stream, the lymph and the nodes and vessels that transport it including the Immune system: defending against disease-causing agents with leukocytes, tonsils, adenoids, thymus and spleen.

- Integumentary system: skin, hair and nails of mammals. Also scales of fish, reptiles, and birds, and feathers of birds.

- Muscular system: movement with muscles.

- Nervous system: collecting, transferring and processing information with brain, spinal cord and nerves.

- Reproductive system: the sex organs, such as ovaries, fallopian tubes, uterus, vulva, vagina, testes, vas deferens, seminal vesicles, prostate and penis.

- Respiratory system: the organs used for breathing, the pharynx, larynx, trachea, bronchi, lungs and diaphragm.

- Skeletal system: structural support and protection with bones, cartilage, ligaments and tendons

Organ Systems

A group of related organs is an *organ system*. Organs within a system may be related in any number of ways, but relationships of function are most commonly used. For example, the urinary system comprises organs that work together to produce, store, and carry urine.

The functions of organ systems often share significant overlap. For instance, the nervous and endocrine system both operate via a shared organ, the hypothalamus. For this reason, the two systems often are combined and studied as the neuroendocrine system. The same is true for the musculoskeletal system, which involves the relationship between the muscular and skeletal systems.

Organ systems as a model for human society

The harmonious and hierarchical bi-level functionality manifested throughout the levels of organization within the organism—from cells to tissues, organs, organ systems, and the whole organism—offers a model for relations and organization in human society and civilization. Ideally individuals would contribute to their families (as cells to tissues), their families to their communities and societies (as tissues to organs), their societies to their nations (as organs to organ systems), and their nations to the world (as organ systems to the body), and in turn each would be benefited by those larger entities.

Organs of the Human Body by Region

Head and neck

- Face

- Orbit

- Eye

- Mouth

- Tongue

- Teeth

- Nose

- Ears

- Scalp

- Larynx

- Pharynx

- Salivary glands

- Meninges

- Brain

- Thyroid

- Parathyroid gland

Back and spine

- Vertebra

- Spinal cord

Thorax

- Mammary gland

- Ribs

- Lungs

- Heart

- Mediastinum

- Esophagus

- Diaphragm

Abdomen

- Peritoneum

- Stomach

- Duodenum

- Intestine

- Colon

- Liver

- Spleen

- Pancreas
- Kidney
- Adrenal gland
- Appendix

Pelvis

- Pelvis
- Sacrum
- Coccyx
- Ovaries
- Fallopian tube
- Uterus
- Vagina
- Vulva
- Clitoris
- Perineum
- Urinary bladder
- Testicles
- Rectum
- Penis

Limbs

- Muscle
- Skeleton
- Nerves
- Hand
- Wrist
- Elbow
- Shoulder
- Hip
- Knee
- Ankle

References

- Moore, Keith L.; Dalley, Arthur F.; Agur Anne M. R. (2010). Moore's Clinically Oriented Anatomy. Phildadelphia: Lippincott Williams & Wilkins. pp. 2–3. ISBN 978-1-60547-652-0

- Coffey, J Calvin; O'Leary, D Peter (2016). "The mesentery: structure, function, and role in disease". The Lancet Gastroenterology & Hepatology. 1 (3): 238–247. doi:10.1016/S2468-1253(16)30026-7

- David N., Fredricks (2001). "Microbial Ecology of Human Skin in Health and Disease". Journal of Investigative Dermatology Symposium Proceedings. 6: 167–169. doi:10.1046/j.0022-202x.2001.00039.x. Retrieved 7 February 2017

- Stewart, Andrew (November 1978). "Polykleitos of Argos," One Hundred Greek Sculptors: Their Careers and Extant Works". Journal of Hellenic Studies. 98: 122–131. doi:10.2307/630196. JSTOR 630196

- Moore, Keith L.; Dalley, Arthur F.; Agur Anne M. R. (2010). Moore's Clinically Oriented Anatomy. Phildadelphia: Lippincott Williams & Wilkins. pp. 2–3. ISBN 978-1-60547-652-0

- Garland, Jr, Theodore; Carter, P. A. (1994). "Evolutionary physiology" (PDF). Annual Review of Physiology. 56 (1): 579–621. doi:10.1146/annurev.ph.56.030194.003051. PMID 8010752

- Ron Sender; Shai Fuchs; Ron Milo (2016). "Revised estimates for the number of human and bacteria cells in the body". PLOS Biology. 14 (8): e1002533. bioRxiv 036103. doi:10.1371/journal.pbio.1002533. PMC 4991899. PMID 27541692

- Zimmer, Carl (2004). "Soul Made Flesh: The Discovery of the Brain – and How It Changed the World". J Clin Invest. 114 (5): 604–04. doi:10.1172/JCI22882. PMC 514597

Chapter 2

Systems of the Human Body

The smooth functioning of the human body is achieved through different kinds of cells, tissues and organ systems. The primary organ systems fundamental to the working of the human body are the cardiovascular system, nervous system, digestive system, immune system, respiratory system, etc. which have been discussed in elaborate detail in this chapter.

Nervous System

The nervous system consists of the brain, spinal cord, sensory organs, and all of the nerves that connect these organs with the rest of the body. Together, these organs are responsible for the control of the body and communication among its parts. The brain and spinal cord form the control center known as the central nervous system (CNS), where information is evaluated and decisions made. The sensory nerves and sense organs of the peripheral nervous system (PNS) monitor conditions inside and outside of the body and send this information to the CNS. Efferent nerves in the PNS carry signals from the control center to the muscles, glands, and organs to regulate their functions.

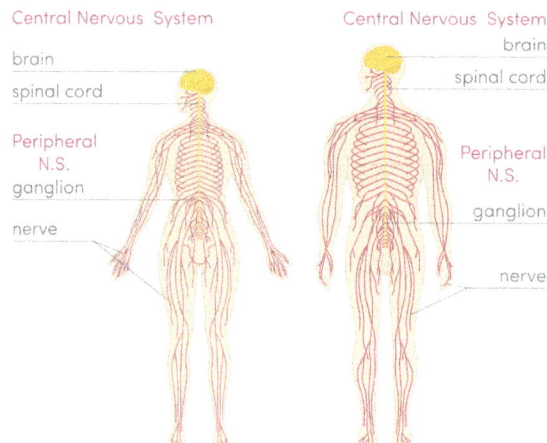

Nervous System Anatomy

Nervous Tissue

The majority of the nervous system is tissue made up of two classes of cells: neurons and neuroglia.

Neurons

Neurons, also known as nerve cells, communicate within the body by transmitting electrochemical signals. Neurons look quite different from other cells in the body due to the many long cellular processes that extend from their central cell body. The cell body is the roughly round part of a neuron that contains the nucleus, mitochondria, and most of the cellular organelles. Small tree-like structures called dendrites extend from the cell body to pick up stimuli from the environment, other neurons, or sensory receptor cells. Long transmitting processes called axons extend from the cell body to send signals onward to other neurons or effector cells in the body.

There are 3 basic classes of neurons: afferent neurons, efferent neurons, and interneurons.

1. *Afferent neurons*: Afferent neurons also known as sensory neurons, afferent neurons transmit sensory signals to the central nervous system from receptors in the body.

2. *Efferent neurons*: Efferent neurons also known as motor neurons, efferent neurons transmit signals from the central nervous system to effectors in the body such as muscles and glands.

3. *Interneurons*: Interneurons form complex networks within the central nervous system to integrate the information received from afferent neurons and to direct the function of the body through efferent neurons.

Neuroglia

Neuroglia, also known as glial cells, act as the "helper" cells of the nervous system. Each neuron in the body is surrounded by anywhere from 6 to 60 neuroglia that protect, feed, and insulate the neuron. Because neurons are extremely specialized cells that are essential to body function and almost never reproduce, neuroglia are vital to maintaining a functional nervous system.

Brain

The brain, a soft, wrinkled organ that weighs about 3 pounds, is located inside the cranial cavity, where the bones of the skull surround and protect it. The approximately 100 billion neurons of the brain form the main control center of the body. The brain and spinal cord together form the central nervous system (CNS), where information is processed and responses originate. The brain, the seat of higher mental functions such as consciousness, memory, planning, and voluntary actions, also controls lower body functions such as the maintenance of respiration, heart rate, blood pressure, and digestion.

Spinal Cord

The spinal cord is a long, thin mass of bundled neurons that carries information through the vertebral cavity of the spine beginning at the medulla oblongata of the brain on its superior end and continuing inferiorly to the lumbar region of the spine. In the lumbar region, the spinal cord separates into a bundle of individual nerves called the cauda equina (due to its resemblance to a horse's tail) that continues inferiorly to the sacrum and coccyx. The white matter of the spinal cord functions as the main conduit of nerve signals to the body from the brain. The grey matter of the spinal cord integrates reflexes to stimuli.

Nerves

Nerves are bundles of axons in the peripheral nervous system (PNS) that act as information highways to carry signals between the brain and spinal cord and the rest of the body. Each axon is wrapped in a connective tissue sheath called the endoneurium. Individual axons of the nerve are bundled into groups of axons called fascicles, wrapped in a sheath of connective tissue called the perineurium. Finally, many fascicles are wrapped together in another layer of connective tissue called the epineurium to form a whole nerve. The wrapping of nerves with connective tissue helps to protect the axons and to increase the speed of their communication within the body.

- *Afferent, Efferent, and Mixed Nerves*: Some of the nerves in the body are specialized for carrying information in only one direction, similar to a one-way street. Nerves that carry information from sensory receptors to the central nervous system only are called afferent nerves. Other neurons, known as efferent nerves, carry signals only from the central nervous system to effectors such as muscles and glands. Finally, some nerves are mixed nerves that contain both afferent and efferent axons. Mixed nerves function like 2-way streets where afferent axons act as lanes heading toward the central nervous system and efferent axons act as lanes heading away from the central nervous system.

- *Cranial Nerves*: Extending from the inferior side of the brain are 12 pairs of cranial nerves. Each cranial nerve pair is identified by a Roman numeral 1 to 12 based upon its location along the anterior-posterior axis of the brain. Each nerve also has a descriptive name (e.g. olfactory, optic, etc.) that identifies its function or location. The cranial nerves provide a direct connection to the brain for the special sense organs, muscles of the head, neck, and shoulders, the heart, and the GI tract.

- *Spinal Nerves*: Extending from the left and right sides of the spinal cord are 31 pairs of spinal nerves. The spinal nerves are mixed nerves that carry both sensory and motor signals between the spinal cord and specific regions of the body. The 31 spinal nerves are split into 5 groups named for the 5 regions of the vertebral column. Thus, there are 8 pairs of cervical nerves, 12 pairs of thoracic nerves, 5 pairs of lumbar nerves, 5 pairs of sacral nerves, and 1 pair of coccygeal nerves. Each spinal nerve exits from the spinal cord through the intervertebral foramen between a pair of vertebrae or between the C1 vertebra and the occipital bone of the skull.

Meninges

The meninges are the protective coverings of the central nervous system (CNS). They consist of three layers: the dura mater, arachnoid mater, and pia mater.

- *Dura mater*: The dura mater, which means "tough mother," is the thickest, toughest, and most superficial layer of meninges. Made of dense irregular connective tissue, it contains many tough collagen fibers and blood vessels. Dura mater protects the CNS from external damage, contains the cerebrospinal fluid that surrounds the CNS, and provides blood to the nervous tissue of the CNS.

- *Arachnoid mater*: The arachnoid mater, which means "spider-like mother," is much thinner and more delicate than the dura mater. It lines the inside of the dura mater and contains many thin fibers that connect it to the underlying pia mater. These fibers cross a fluid-filled space called the subarachnoid space between the arachnoid mater and the pia mater.

- *Pia mater*: The pia mater, which means "tender mother," is a thin and delicate layer of tissue that rests on the outside of the brain and spinal cord. Containing many blood vessels that feed the nervous tissue of the CNS, the pia mater penetrates into the valleys of the sulci and fissures of the brain as it covers the entire surface of the CNS.

Cerebrospinal Fluid

The space surrounding the organs of the CNS is filled with a clear fluid known as cerebrospinal fluid (CSF). CSF is formed from blood plasma by special structures called choroid plexuses. The choroid plexuses contain many capillaries lined with epithelial tissue that filters blood plasma and allows the filtered fluid to enter the space around the brain.

Newly created CSF flows through the inside of the brain in hollow spaces called ventricles and through a small cavity in the middle of the spinal cord called the central canal. CSF also flows through the subarachnoid space around the outside of the brain and spinal cord. CSF is constantly produced at the choroid plexuses and is reabsorbed into the bloodstream at structures called arachnoid villi.

Cerebrospinal fluid provides several vital functions to the central nervous system:

1. CSF absorbs shocks between the brain and skull and between the spinal cord and vertebrae. This shock absorption protects the CNS from blows or sudden changes in velocity, such as during a car accident.

2. The brain and spinal cord float within the CSF, reducing their apparent weight through buoyancy. The brain is a very large but soft organ that requires a high volume of blood to function effectively. The reduced weight in cerebrospinal fluid allows the blood vessels of the brain to remain open and helps protect the nervous tissue from becoming crushed under its own weight.

3. CSF helps to maintain chemical homeostasis within the central nervous system. It contains ions, nutrients, oxygen, and albumins that support the chemical and osmotic balance of nervous tissue. CSF also removes waste products that form as byproducts of cellular metabolism within nervous tissue.

Sense Organs

All of the bodies' many sense organs are components of the nervous system. What are known as the special senses—vision, taste, smell, hearing, and balance—are all detected by specialized organs such as the eyes, taste buds, and olfactory epithelium. Sensory receptors for the general senses like touch, temperature, and pain are found throughout most of the body. All of the sensory receptors of the body are connected to afferent neurons that carry their sensory information to the CNS to be processed and integrated.

Nervous System Physiology

Functions of the Nervous System

The nervous system has 3 main functions: sensory, integration, and motor.

1. *Sensory*: The sensory function of the nervous system involves collecting information from sensory receptors that monitor the body's internal and external conditions. These signals are then passed on to the central nervous system (CNS) for further processing by afferent neurons (and nerves).

2. *Integration*: The process of integration is the processing of the many sensory signals that are passed into the CNS at any given time. These signals are evaluated, compared, used for decision making, discarded or committed to memory as deemed appropriate. Integration takes place in the gray matter of the brain and spinal cord and is performed by interneurons. Many interneurons work together to form complex networks that provide this processing power.

3. *Motor*: Once the networks of interneurons in the CNS evaluate sensory information and decide on an action, they stimulate efferent neurons. Efferent neurons (also called motor neurons) carry signals from the gray matter of the CNS through the nerves of the peripheral nervous system to effector cells. The effector may be smooth, cardiac, or skeletal muscle tissue or glandular tissue. The effector then releases a hormone or moves a part of the body to respond to the stimulus.

Unfortunately of course, our nervous system doesn't always function as it should. Sometimes this is the result of diseases like Alzheimer's and Parkinson's disease.

Divisions of the Nervous System

Central Nervous System

The brain and spinal cord together form the central nervous system, or CNS. The CNS acts as the control center of the body by providing its processing, memory, and regulation systems. The CNS takes in all of the conscious and subconscious sensory information from the body's sensory receptors to stay aware of the body's internal and external conditions. Using this sensory information, it makes decisions about both conscious and subconscious actions to take to maintain the body's homeostasis and ensure its survival. The CNS is also responsible for the higher functions of the nervous system such as language, creativity, expression, emotions, and personality. The brain is the seat of consciousness and determines who we are as individuals.

Peripheral Nervous System

The peripheral nervous system (PNS) includes all of the parts of the nervous system outside of the brain and spinal cord. These parts include all of the cranial and spinal nerves, ganglia, and sensory receptors.

Somatic Nervous System

The somatic nervous system (SNS) is a division of the PNS that includes all of the voluntary efferent neurons. The SNS is the only consciously controlled part of the PNS and is responsible for stimulating skeletal muscles in the body.

Autonomic Nervous System

The autonomic nervous system (ANS) is a division of the PNS that includes all of the involuntary efferent neurons. The ANS controls subconscious effectors such as visceral muscle tissue, cardiac muscle tissue, and glandular tissue.

There are 2 divisions of the autonomic nervous system in the body: the sympathetic and parasympathetic divisions.

- *Sympathetic*: The sympathetic division forms the body's "fight or flight" response to stress, danger, excitement, exercise, emotions, and embarrassment. The sympathetic division increases respiration and heart rate, releases adrenaline and other stress hormones, and decreases digestion to cope with these situations.

- *Parasympathetic*: The parasympathetic division forms the body's "rest and digest" response when the body is relaxed, resting, or feeding. The parasympathetic works to undo the work of the sympathetic division after a stressful situation. Among other functions, the parasympathetic division works to decrease respiration and heart rate, increase digestion, and permit the elimination of wastes.

Enteric Nervous System

The enteric nervous system (ENS) is the division of the ANS that is responsible for regulating digestion and the function of the digestive organs. The ENS receives signals from the central nervous system through both the sympathetic and parasympathetic divisions of the autonomic nervous system to help regulate its functions. However, the ENS mostly works independently of the CNS and continues to function without any outside input. For this reason, the ENS is often called the "brain of the gut" or the body's "second brain." The ENS is an immense system—almost as many neurons exist in the ENS as in the spinal cord.

Action Potentials

Neurons function through the generation and propagation of electrochemical signals known as action potentials (APs). An AP is created by the movement of sodium and potassium ions through the membrane of neurons.

- *Resting Potential*: At rest, neurons maintain a concentration of sodium ions outside of the cell and potassium ions inside of the cell. This concentration is maintained by the sodium-potassium pump of the cell membrane which pumps 3 sodium ions out of the cell for every 2 potassium ions that are pumped into the cell. The ion concentration results in a resting electrical potential of -70 millivolts (mV), which means that the inside of the cell has a negative charge compared to its surroundings.

- *Threshold Potential*: If a stimulus permits enough positive ions to enter a region of the cell to cause it to reach -55 mV, that region of the cell will open its voltage-gated sodium channels and allow sodium ions to diffuse into the cell. -55 mV is the threshold potential for neurons as this is the "trigger" voltage that they must reach to cross the threshold into forming an action potential.

- *Depolarization*: Sodium carries a positive charge that causes the cell to become depolarized (positively charged) compared to its normal negative charge. The voltage for depolarization of all neurons is +30 mV. The depolarization of the cell is the AP that is transmitted by the neuron as a nerve signal. The positive ions spread into neighboring regions of the cell, initiating a new AP in those regions as they reach -55 mV. The AP continues to spread down the cell membrane of the neuron until it reaches the end of an axon.

- *Repolarization*: After the depolarization voltage of +30 mV is reached, voltage-gated potassium ion channels open, allowing positive potassium ions to diffuse out of the cell. The loss of potassium along with the pumping of sodium ions back out of the cell through the sodium-potassium pump restores the cell to the -55 mV resting potential. At this point the neuron is ready to start a new action potential.

Synapses

A synapse is the junction between a neuron and another cell. Synapses may form between 2 neurons or between a neuron and an effector cell. There are two types of synapses found in the body: chemical synapses and electrical synapses.

- *Chemical synapses*: At the end of a neuron's axon is an enlarged region of the axon known as the axon terminal. The axon terminal is separated from the next cell by a small gap known as the synaptic cleft. When an AP reaches the axon terminal, it opens voltage-gated calcium ion channels. Calcium ions cause vesicles containing chemicals known as neurotransmitters (NT) to release their contents by exocytosis into the synaptic cleft. The NT molecules cross the synaptic cleft and bind to receptor molecules on the cell, forming a synapse with the neuron. These receptor molecules open ion channels that may either stimulate the receptor cell to form a new action potential or may inhibit the cell from forming an action potential when stimulated by another neuron.

- *Electrical synapses*: Electrical synapses are formed when 2 neurons are connected by small holes called gap junctions. The gap junctions allow electric current to pass from one neuron to the other, so that an AP in one cell is passed directly on to the other cell through the synapse.

Myelination

The axons of many neurons are covered by a coating of insulation known as myelin to increase the speed of nerve conduction throughout the body. Myelin is formed by 2 types of glial cells: Schwann cells in the PNS and oligodendrocytes in the CNS. In both cases, the glial cells wrap their plasma membrane around the axon many times to form a thick covering of lipids. The development of these myelin sheaths is known as myelination.

Myelination speeds up the movement of APs in the axon by reducing the number of APs that must form for a signal to reach the end of an axon. The myelination process begins speeding up nerve conduction in fetal development and continues into early adulthood. Myelinated axons appear white due to the presence of lipids and form the white matter of the inner brain and outer spinal cord. White matter is specialized for carrying information quickly through the brain and spinal cord. The gray matter of the brain and spinal cord are the unmyelinated integration centers where information is processed.

Reflexes

Reflexes are fast, involuntary responses to stimuli. The most well known reflex is the patellar reflex, which is checked when a physicians taps on a patient's knee during a physical examination. Reflexes are integrated in the gray matter of the spinal cord or in the brain stem. Reflexes allow the body to respond to stimuli very quickly by sending responses to effectors before the nerve signals reach the conscious parts of the brain. This explains why people will often pull their hands away from a hot object before they realize they are in pain.

Functions of the Cranial Nerves

Each of the 12 cranial nerves has a specific function within the nervous system.

- The olfactory nerve (I) carries scent information to the brain from the olfactory epithelium in the roof of the nasal cavity.

- The optic nerve (II) carries visual information from the eyes to the brain.

- Oculomotor, trochlear, and abducens nerves (III, IV, and VI) all work together to allow the brain to control the movement and focus of the eyes. The trigeminal nerve (V) carries sensations from the face and innervates the muscles of mastication.

- The facial nerve (VII) innervates the muscles of the face to make facial expressions and carries taste information from the anterior 2/3 of the tongue.

- The vestibulocochlear nerve (VIII) conducts auditory and balance information from the ears to the brain.

- The glossopharyngeal nerve (IX) carries taste information from the posterior 1/3 of the tongue and assists in swallowing.

- The vagus nerve (X), sometimes called the wandering nerve due to the fact that it innervates many different areas, "wanders" through the head, neck, and torso. It carries information about the condition of the vital organs to the brain, delivers motor signals to control speech and delivers parasympathetic signals to many organs.

- The accessory nerve (XI) controls the movements of the shoulders and neck.

- The hypoglossal nerve (XII) moves the tongue for speech and swallowing.

Sensory Physiology

All sensory receptors can be classified by their structure and by the type of stimulus that they detect. Structurally, there are 3 classes of sensory receptors: free nerve endings, encapsulated nerve endings, and specialized cells. Free nerve endings are simply free dendrites at the end of a neuron that extend into a tissue. Pain, heat, and cold are all sensed through free nerve endings. An encapsulated nerve ending is a free nerve ending wrapped in a round capsule of connective tissue. When the capsule is deformed by touch or pressure, the neuron is stimulated to send signals to the CNS. Specialized cells detect stimuli from the 5 special senses: vision, hearing, balance, smell, and taste. Each of the special senses has its own unique sensory cells—such as rods and cones in the retina to detect light for the sense of vision.

Functionally, there are 6 major classes of receptors: mechanoreceptors, nociceptors, photoreceptors, chemoreceptors, osmoreceptors, and thermoreceptors.

- *Mechanoreceptors*: Mechanoreceptors are sensitive to mechanical stimuli like touch, pressure, vibration, and blood pressure.

- *Nociceptors*: Nociceptors respond to stimuli such as extreme heat, cold, or tissue damage by sending pain signals to the CNS.

- *Photoreceptors*: Photoreceptors in the retina detect light to provide the sense of vision.

- *Chemoreceptors*: Chemoreceptors detect chemicals in the bloodstream and provide the senses of taste and smell.

- *Osmoreceptors*: Osmoreceptors monitor the osmolarity of the blood to determine the body's hydration levels.

- *Thermoreceptors*: Thermoreceptors detect temperatures inside the body and in its surroundings.

Central Nervous System

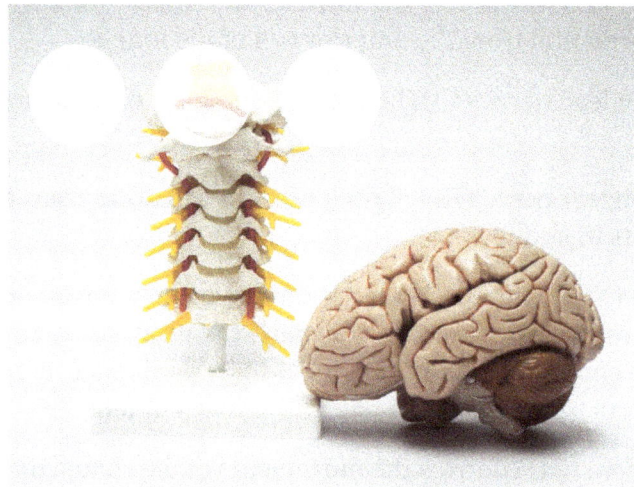

The CNS is the brain and spinal cord

The CNS consists of the brain and spinal cord.

The brain is protected by the skull (the cranial cavity) and the spinal cord travels from the back of the brain, down the center of the spine, stopping in the lumbar region of the lower back.

The brain and spinal cord are both housed within a protective triple-layered membrane called the meninges.

The central nervous system has been thoroughly studied by anatomists and physiologists, but it still holds many secrets; it controls our thoughts, movements, emotions, and desires. It also controls our breathing, heart rate, the release of some hormones, body temperature, and much more.

The retina, optic nerve, olfactory nerves, and olfactory epithelium are sometimes considered to be part of the CNS alongside the brain and spinal cord. This is because they connect directly with brain tissue without intermediate nerve fibers.

Different Parts of the CNS

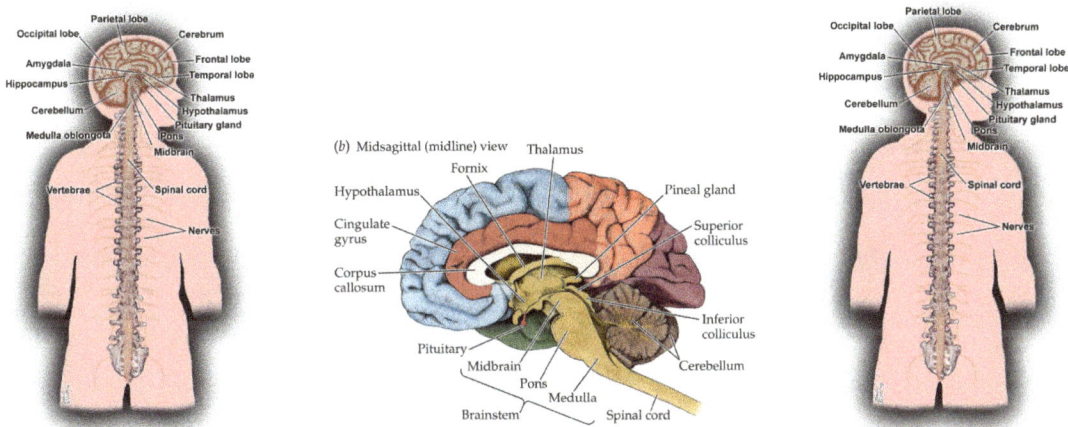

Picture of the central nervous system (CNS)

The Brain and Cerebrum

The cerebrum is the largest part of the brain and controls voluntary actions, speech, senses, thought, and memory.

The surface of the cerebral cortex has grooves or infoldings (called sulci), the largest of which are termed fissures. Some fissures separate lobes.

The convolutions of the cortex give it a wormy appearance. Each convolution is delimited by two sulci and is also called a gyrus (gyri in plural). The cerebrum is divided into two halves, known as the right and left hemispheres. A mass of fibers called the corpus callosum links the hemispheres. The right hemisphere controls voluntary limb movements on the left side of the body, and the left hemisphere controls voluntary limb movements on the right side of the body. Almost every person has one dominant hemisphere. Each hemisphere is divided into four lobes, or areas, which are interconnected.

- The frontal lobes are located in the front of the brain and are responsible for voluntary movement and, via their connections with other lobes, participate in the execution of sequential tasks; speech output; organizational skills; and certain aspects of behavior, mood, and memory.

- The parietal lobes are located behind the frontal lobes and in front of the occipital lobes. They process sensory information such as temperature, pain, taste, and touch. In addition, the processing includes information about numbers, attentiveness to the position of one's body parts, the space around one's body, and one's relationship to this space.

- The temporal lobes are located on each side of the brain. They process memory and auditory (hearing) information and speech and language functions.

- The occipital lobes are located at the back of the brain. They receive and process visual information.

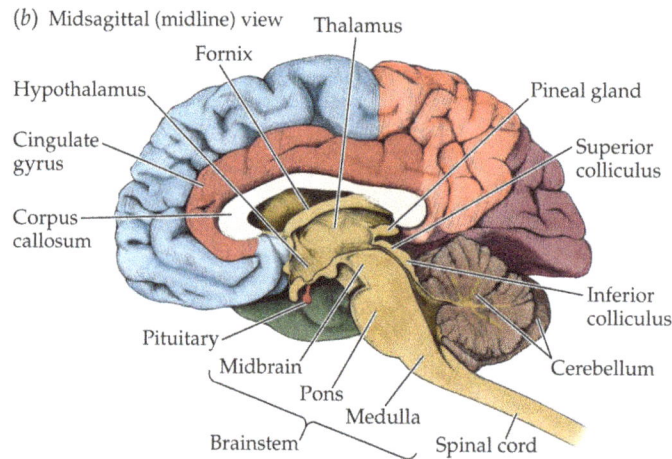

(b) Midsagittal (midline) view

Anatomy of the brain

The cortex, also called gray matter, is the most external layer of the brain and predominantly contains neuronal bodies (the part of the neurons where the DNA-containing cell nucleus is located). The gray matter participates actively in the storage and processing of information. An isolated clump of nerve cell bodies in the gray matter is termed a nucleus (to be differentiated from a cell nucleus). The cells in the gray matter extend their projections, called axons, to other areas of the brain.

Fibers that leave the cortex to conduct impulses toward other areas are termed efferent fibers, and fibers that approach the cortex from other areas of the nervous system are termed afferent (nerves or pathways). Fibers that go from the motor cortex to the brainstem (for example, the pons) or the spinal cord receive a name that generally reflects the connections (that is, corticopontine tract for the former and corticospinal tract for the latter). Axons are surrounded in their course outside the gray matter by myelin, which has a glistening whitish appearance and thus gives rise to the term white matter.

Cortical areas receive their names according to their general function or lobe name. If in charge of motor function, the area is called the motor cortex. If in charge of sensory function, the area is called a sensory or somesthetic cortex. The calcarine or visual cortex is located in the occipital lobe (also termed occipital cortex) and receives visual input. The auditory cortex, localized in the temporal lobe, processes sounds or verbal input. Knowledge of the anatomical projection of fibers of the different tracts and the relative representation of body regions in the cortex often enables doctors to correctly locate an injury and its relative size, sometimes with great precision.

The Central Structures of the Brain

The central structures of the brain include the thalamus, hypothalamus, and pituitary gland. The hippocampus is located in the temporal lobe but participates in the processing of memory and emotions and is interconnected with central structures. Other structures are the basal ganglia, which are made up of gray matter and include the amygdala (localized in the temporal lobe), the caudate nucleus, and the lenticular nucleus (putamen and globus pallidus). Because the caudate and putamen are structurally similar, neuropathologists have coined for them the collective term striatum.

- The thalamus integrates and relays sensory information to the cortex of the parietal, temporal, and occipital lobes. The thalamus is located in the lower central part of the brain (that is, upper part of the brainstem) and is located medially to the basal ganglia. The brain hemispheres lie on the thalamus. Other roles of the thalamus include motor and memory control.

- The hypothalamus, located below the thalamus, regulates automatic functions such as appetite, thirst, and body temperature. It also secretes hormones that stimulate or suppress the release of hormones (for example, growth hormones) in the pituitary gland.

- The pituitary gland is located at the base of the brain. The pituitary gland produces hormones that control many functions of other endocrine glands. It regulates the production of many hormones that have a role in growth, metabolism, sexual response, fluid and mineral balance, and the stress response.

- The ventricles are cerebrospinal fluid-filled cavities in the interior of the cerebral hemispheres.

The Base of the Brain

The base of the brain contains the cerebellum and the brainstem. These structures serve complex functions. Below is a simplified version of these roles:

- Traditionally, the cerebellum has been known to control equilibrium and coordination and contributes to the generation of muscle tone. It has more recently become evident, however, that the cerebellum plays more diverse roles such as participating in some types of memory and exerting a complex influence on musical and mathematical skills.

- The brainstem connects the brain with the spinal cord. It includes the midbrain, the pons, and the medulla oblongata. It is a compact structure in which multiple pathways traverse from the brain to the spinal cord and vice versa. For instance, nerves that arise from cranial nerve nuclei are involved with eye movements and exit the brainstem at several levels. Damage to the brainstem can therefore affect a number of bodily functions. For instance, if the corticospinal tract is injured, a loss of motor function (paralysis) occurs, and it may be accompanied by other neurologic deficits, such as eye movement abnormalities, which are reflective of injury to cranial nerves or their pathways in the brainstem.

  - The midbrain is located below the hypothalamus. Some cranial nerves that are also responsible for eye muscle control exit the midbrain.

- The pons serves as a bridge between the midbrain and the medulla oblongata. The pons also contains the nuclei and fibers of nerves that serve eye muscle control, facial muscle strength, and other functions.

- The medulla oblongata is the lowest part of the brainstem and is interconnected with the cervical spinal cord. The medulla oblongata also helps control involuntary actions, including vital processes, such as heart rate, blood pressure, and respiration, and it carries the corticospinal (that is, motor function) tract toward the spinal cord.

Multiple Sclerosis

Multiple Sclerosis or MS is an autoimmune disorder that happens when your body's immune system attacks the covering (myelin sheath) that surrounds the nerves of the central nervous system (CNS). Doctors and researchers don't know exactly a person gets MS, but they believe it's related to genetics, acquired (you get it from an infection, etc.), or from the environment.

The Spinal Cord

The spinal cord is an extension of the brain and is surrounded by the vertebral bodies that form the spinal column. The central structures of the spinal cord are made up of gray matter (nerve cell bodies), and the external or surrounding tissues are made up of white matter.

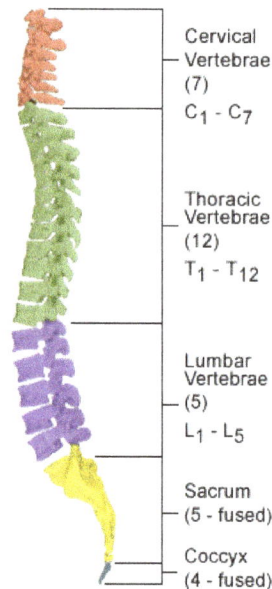

Cervical Vertebrae (7) $C_1 - C_7$

Thoracic Vertebrae (12) $T_1 - T_{12}$

Lumbar Vertebrae (5) $L_1 - L_5$

Sacrum (5 - fused)

Coccyx (4 - fused)

The anatomy of the spine

Within the spinal cord are 30 segments that belong to 4 sections (cervical, thoracic, lumbar, sacral), based on their location:

- Eight cervical segments: These transmit signals from or to areas of the head, neck, shoulders, arms, and hands.

- Twelve thoracic segments: These transmit signals from or to part of the arms and the anterior and posterior chest and abdominal areas.

- Five lumbar segments: These transmit signals from or to the legs and feet and some pelvic organs.

- Five sacral segments: These transmit signals from or to the lower back and buttocks, pelvic organs and genital areas, and some areas in the legs and feet.

- A coccygeal remnant is located at the bottom of the spinal cord.

Peripheral Nervous System

The peripheral nervous system (PNS) consists of all neurons that exist outside the brain and spinal cord. This includes long nerve fibers containing bundles of axons as well as ganglia made of neural cell bodies. The peripheral nervous system connects the central nervous system (CNS) made of the brain and spinal cord to various parts of the body and receives input from the external environment as well.

Functionally, the PNS is divided into sensory (afferent) and motor (efferent) nerves, depending on whether they bring information to the CNS from sensory receptors or carry instructions towards muscles, organs or other effectors. Motor nerves can be further classified as somatic or autonomic nerves, depending on whether the motor activity is under voluntary conscious control.

Anatomically, the PNS can be divided into spinal and cranial nerves, depending on whether they emerge from the spinal cord or the brain and brainstem. Both cranial and spinal nerves can have sensory, motor or mixed functions. The enteric nervous system, surrounding the gastrointestinal tract is another important part of the peripheral nervous system. While it receives signals from the autonomic nervous system, it can function independently as well and contains nearly five times as many neurons as the spinal cord.

Functions of the Peripheral Nervous System

The primary function of the peripheral nervous system is to connect the brain and spinal cord to the rest of the body and the external environment. This is accomplished through nerves that carry information from sensory receptors in the eyes, ears, skin, nose and tongue, as well as stretch receptors and nociceptors in muscles, glands and other internal organs. When the CNS integrates these varied signals, and formulates a response, motor nerves of the PNS innervate effector organs and mediate the contraction or relaxation of skeletal, smooth or cardiac muscle.

Thus, the PNS regulates internal homeostasis through the autonomic nervous system, modulating respiration, heart rate, blood pressure, digestion reproduction and immune responses. It can increase or decrease the strength of muscle contractility across the body, whether it is sphincters in the digestive and excretory systems, cardiac muscles in the heart or skeletal muscles for movement. It is necessary for all voluntary action, balance and maintenance of posture and for the release of secretions from most exocrine glands. The PNS innervates the muscles surrounding sense organs, so it is involved in chewing, swallowing, biting and speaking. At the same time, it mediates the response of the body to noxious stimuli, quickly removing the body from the injurious stimulus, whether it is extremes in temperature, pH, or pressure, as well as stretching and compressing forces.

Examples of the Peripheral Nervous System Response

In a dimly lit room, the pupils of the eye are enlarged, to allow maximum light to fall on the retina. When a bright light is suddenly turned on, sensory receptors in the eye communicate this to the CNS. The response to this new stimulus is mediated through the peripheral nervous system, by contracting the pupils, using external eye muscles to squint and probably even moving the skeletal muscles of the arm to shield the eye.

Similarly, when a sharp or pointed object is stepped on, pain and stretch receptors in the skin send signals to the CNS, which immediately brings about a change in posture, and balance, protecting the foot from potential injury.

Anatomy of the Peripheral Nervous System

The peripheral nervous system is made of nerves, ganglia and plexuses. A nerve contains the axons of multiple neurons bound together by connective tissue. The axon itself is often myelinated, containing a phospholipid secreted by a glial cell called the Schwann cell. The thin covering of Schwann cell cytoplasm forms the innermost layer protecting an axon and is called the neurilemma or neurolemma.

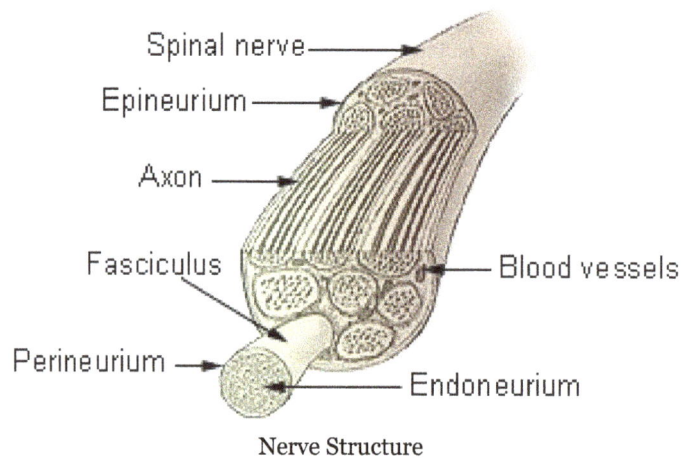

Nerve Structure

The image above depicts the structure of a nerve. Blood capillaries and other connective tissue around the neurilemma form the endoneurium. When multiple axons are bundled together to form structures called fascicles, a fibrous tissue called the perineurium holds them together. Finally, the whole nerve containing numerous axon bundles is encased in fibrous epineurium. The cell bodies or soma of these neurons also cluster together and are covered by the epineurium to form ganglia that look like swellings on the nerve fiber. In the autonomic nervous system, these ganglia become the sites for synaptic transmission between two neurons. Branching networks of intersecting spinal and autonomic nerves form structures called plexuses that have both sensory and motor functions and serve a particular region of the body.

The PNS can be said to consist of 12 pairs of cranial nerves and 31 pairs of spinal nerves. Cranial nerves emerge in pairs on either side of the base of the skull, through small openings called foramina. Cranial nerves are numbered using roman numerals I-XII, depending on their position while exiting the cranium. A potentially vestigial nerve called cranial nerve zero emerges anterior to the first cranial nerve. Cranial nerves also have a Latin or Greek name, based on their structure or their effector organ. They primarily innervate the head and neck, with the significant exception of the tenth cranial nerve, also known as the vagus nerve. Some cranial nerves have only sensory function, such as the olfactory and optic nerves. The structure of these nerves also occasionally leads to their classification under the central nervous system. Cranial nerve VIII is another sensory nerve relating to hearing and balance. Motor nerves contain nerve fibers that carry signals to muscles of the pupil of the eye or external eye muscles. The rest are mixed nerves containing both sensory and motor nerve fibers. Among these, the XI and XII cranial nerves mostly serve a motor function, and innervate the neck, back and tongue. The Vagus nerve is another mixed nerve that carries signals from internal organs to the brain and conducts impulses to the organs of the thorax, abdomen and respiratory muscles of the pharynx and larynx. It plays an important role in the parasympathetic innervation of the body.

There are 31 pairs of spinal nerves, arising from different regions of the spinal cord. There are 8 that emerge from the cervical region, 12 from the thoracic region, 5 each from the lumbar and sacral regions and 1 pair of spinal nerves from the coccygeal region. Each spinal nerve is a mixed nerve formed by a combination of afferent and efferent neurons.

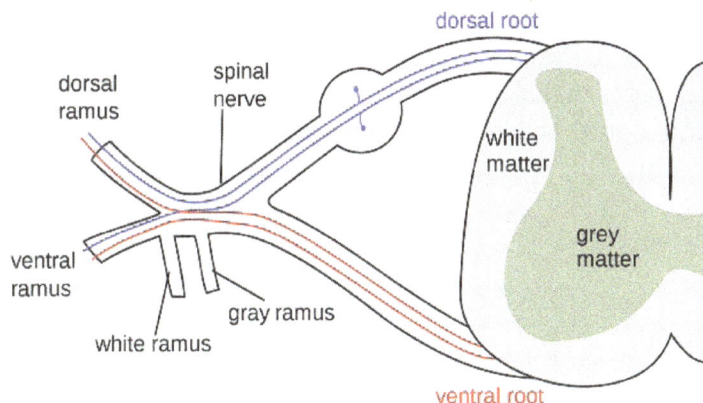

The image shows the region near the spinal cord, where every spinal nerve has a posterior and anterior root. The anterior or ventral root contains motor neurons while the dorsal or posterior root has ganglia containing the cell bodies of afferent sensory neurons. Distal to the spine, the nerve again splits into an anterior and posterior ramus in addition to forming a small meningeal branch. The posterior ramus leads to the muscles, joints and skin on the back. The anterior ramus is involved in innervating the skin and muscles of the trunk and leads towards the limbs. Often, the anterior ramus forms a network of intersecting nerve fibers to create plexuses.

Sensory Nervous System

The functional classification of the PNS divides it into three categories. The first is the sensory nervous system, carrying signals from the viscera, sense organs, muscles, bones and joints towards the CNS. Nerve fibers that carry this information are part of the afferent division. Sensory receptors can transduce a physical stimulus such as pressure, sound waves, electromagnetic radiation, or chemical composition, into an electrochemical signal. This signal, when it reaches a certain threshold, is transmitted as an action potential along an afferent neuron, and relayed to the CNS, where the signal is perceived and interpreted. Thus the sensory nervous system consisting of the receptor and neural pathway deliver information about intensity, location, type and duration of a stimulus to the CNS.

Somatic Nervous System

The second functional division of the PNS is the somatic nervous system. It controls voluntary muscular movement of skeletal muscles in the limbs, back, shoulders, neck and face. It also mediates reflex actions, where an afferent nerve fiber is nearly directly connected to a motor nerve fiber, to quickly generate a response to stimulus. These include protective responses, like the movement of the body away from acute injurious stimuli like extremes in temperature, as well as those like the patellar 'knee-jerk' response when the patellar ligament is struck.

Autonomic Nervous System

The autonomic nervous system is related to all the involuntary visceral activity of the body. It consists of the sympathetic and parasympathetic nervous systems, and their effector organs include cardiac muscle, smooth muscle, and various glands. The anatomy of the autonomic nervous system is distinct because the effector arm involves two neurons that synapse with each other at specific ganglia. The neurons of the sympathetic nervous system have short preganglionic neurons that can excite multiple postganglionic nerve fibers. The sympathetic nervous system is said to have a thoracic and lumbar outflow. The parasympathetic nervous system, on the other hand uses cranial and sacral nerves and their ganglia are situated close to the target organ.

Related Biology Terms

- Afferent Nerve – Axon of an afferent sensory neuron that carries signals towards the central nervous system.

- Cranium – Bony structure present at the top of the vertebral column, protecting the brain and major sense organs.

- Neurolemma – Outermost thin layer of nucleated Schwann cells' cytoplasm that surrounds the axons of peripheral nerves. Serves to protect the neuron and helps in recovery after injury.

- Nociceptor – Sensory nerve cell that is activated in response to a potentially injurious stimulus and creates the sensation of pain.

Muscular System

The muscular system is responsible for the movement of the human body. Attached to the bones of the skeletal system are about 700 named muscles that make up roughly half of a person's body weight. Each of these muscles is a discrete organ constructed of skeletal muscle tissue, blood vessels, tendons, and nerves. Muscle tissue is also found inside of the heart, digestive organs, and blood vessels. In these organs, muscles serve to move substances throughout the body.

Muscular System Anatomy

Muscle Types

There are three types of muscle tissue: Visceral, cardiac, and skeletal.

Visceral Muscle

Visceral muscle is found inside of organs like the stomach, intestines, and blood vessels. The weakest of all muscle tissues, visceral muscle makes organs contract to move substances through the organ. Because visceral muscle is controlled by the unconscious part of the brain, it is known as involuntary muscle—it cannot be directly controlled by the conscious mind. The term "smooth muscle" is often used to describe visceral muscle because it has a very smooth, uniform appearance when viewed under a microscope. This smooth appearance starkly contrasts with the banded appearance of cardiac and skeletal muscles.

Cardiac Muscle

Found only in the heart, cardiac muscle is responsible for pumping blood throughout the body. Cardiac muscle tissue cannot be controlled consciously, so it is an involuntary muscle. While hormones and signals from the brain adjust the rate of contraction, cardiac muscle stimulates itself to contract. The natural pacemaker of the heart is made of cardiac muscle tissue that stimulates other cardiac muscle cells to contract. Because of its self-stimulation, cardiac muscle is considered to be autorhythmic or intrinsically controlled.

The cells of cardiac muscle tissue are striated—that is, they appear to have light and dark stripes when viewed under a light microscope. The arrangement of protein fibers inside of the cells causes these light and dark bands. Striations indicate that a muscle cell is very strong, unlike visceral muscles.

The cells of cardiac muscle are branched X or Y shaped cells tightly connected together by special junctions called intercalated disks. Intercalated disks are made up of fingerlike projections from two neighboring cells that interlock and provide a strong bond between the cells. The branched structure and intercalated disks allow the muscle cells to resist high blood pressures and the strain of pumping blood throughout a lifetime. These features also help to spread electrochemical signals quickly from cell to cell so that the heart can beat as a unit.

Skeletal Muscle

Skeletal muscle is the only voluntary muscle tissue in the human body—it is controlled consciously. Every physical action that a person consciously performs (e.g. speaking, walking, or writing) requires skeletal muscle. The function of skeletal muscle is to contract to move parts of the body closer to the bone that the muscle is attached to. Most skeletal muscles are attached to two bones across a joint, so the muscle serves to move parts of those bones closer to each other.

Skeletal muscle cells form when many smaller progenitor cells lump themselves together to form long, straight, multinucleated fibers. Striated just like cardiac muscle, these skeletal muscle fibers are very strong. Skeletal muscle derives its name from the fact that these muscles always connect to the skeleton in at least one place.

Gross Anatomy of a Skeletal Muscle

Most skeletal muscles are attached to two bones through tendons. Tendons are tough bands of dense regular connective tissue whose strong collagen fibers firmly attach muscles to bones. Tendons are under extreme stress when muscles pull on them, so they are very strong and are woven into the coverings of both muscles and bones.

Muscles move by shortening their length, pulling on tendons, and moving bones closer to each other. One of the bones is pulled towards the other bone, which remains stationary. The place on the stationary bone that is connected via tendons to the muscle is called the origin. The place on the moving bone that is connected to the muscle via tendons is called the insertion. The belly of the muscle is the fleshy part of the muscle in between the tendons that does the actual contraction.

Names of Skeletal Muscles

Skeletal muscles are named based on many different factors, including their location, origin and insertion, number of origins, shape, size, direction, and function.

- *Location*: Many muscles derive their names from their anatomical region. The rectus abdominis and transverse abdominis, for example, are found in the abdominal region. Some muscles, like the tibialis anterior, are named after the part of the bone (the anterior portion of the tibia) that they are attached to. Other muscles use a hybrid of these two, like the brachioradialis, which is named after a region (brachial) and a bone (radius).

- *Origin and Insertion*: Some muscles are named based upon their connection to a stationary bone (origin) and a moving bone (insertion). These muscles become very easy to identify once you know the names of the bones that they are attached to. Examples of this type of muscle include the sternocleidomastoid (connecting the sternum and clavicle to the mastoid process of the skull) and the occipitofrontalis (connecting the occipital bone to the frontal bone).

- *Number of Origins*: Some muscles connect to more than one bone or to more than one place on a bone, and therefore have more than one origin. A muscle with two origins is called a biceps. A muscle with three origins is a triceps muscle. Finally, a muscle with four origins is a quadriceps muscle.

- *Shape, Size, and Direction*: We also classify muscles by their shapes. For example, the deltoids have a delta or triangular shape. The serratus muscles feature a serrated or saw-like shape. The rhomboid major is a rhombus or diamond shape. The size of the muscle can be used to distinguish between two muscles found in the same region. The gluteal region contains three muscles differentiated by size—the gluteus maximus (large), gluteus medius (medium), and gluteus minimus (smallest). Finally, the direction in which the muscle fibers run can be used to identify a muscle. In the abdominal region, there are several sets of wide, flat muscles. The muscles whose fibers run straight up and down are the rectus abdominis, the ones running transversely (left to right) are the transverse abdominis, and the ones running at an angle are the obliques.

- *Function*: Muscles are sometimes classified by the type of function that they perform. Most of the muscles of the forearms are named based on their function because they are located in the same region and have similar shapes and sizes. For example, the flexor group of the forearm flexes the wrist and the fingers. The supinator is a muscle that supinates the wrist by rolling it over to face palm up. In the leg, there are muscles called adductors whose role is to adduct (pull together) the legs.

Groups Action in Skeletal Muscle

Skeletal muscles rarely work by themselves to achieve movements in the body. More often they work in groups to produce precise movements. The muscle that produces any particular movement of the body is known as an agonist or prime mover. The agonist always pairs with an antagonist muscle that produces the opposite effect on the same bones. For example, the biceps brachii muscle flexes the arm at the elbow. As the antagonist for this motion, the triceps brachii muscle extends the arm at the elbow. When the triceps is extending the arm, the biceps would be considered the antagonist.

In addition to the agonist/antagonist pairing, other muscles work to support the movements of the agonist. Synergists are muscles that help to stabilize a movement and reduce extraneous movements. They are usually found in regions near the agonist and often connect to the same bones. Because skeletal muscles move the insertion closer to the immobile origin, fixator muscles assist in movement by holding the origin stable. If you lift something heavy with your arms, fixators in the trunk region hold your body upright and immobile so that you maintain your balance while lifting.

Skeletal Muscle Histology

Skeletal muscle fibers differ dramatically from other tissues of the body due to their highly specialized functions. Many of the organelles that make up muscle fibers are unique to this type of cell.

The sarcolemma is the cell membrane of muscle fibers. The sarcolemma acts as a conductor for electrochemical signals that stimulate muscle cells. Connected to the sarcolemma are transverse tubules (T-tubules) that help carry these electrochemical signals into the middle of the muscle fiber. The sarcoplasmic reticulum serves as a storage facility for calcium ions ($Ca2+$) that are vital to muscle contraction. Mitochondria, the "power houses" of the cell, are abundant in muscle cells to break down sugars and provide energy in the form of ATP to active muscles. Most of the muscle fiber's structure is made up of myofibrils, which are the contractile structures of the cell. Myofibrils are made up of many proteins fibers arranged into repeating subunits called sarcomeres. The sarcomere is the functional unit of muscle fibers.

Sarcomere Structure

Sarcomeres are made of two types of protein fibers: thick filaments and thin filaments.

- *Thick filaments*. Thick filaments are made of many bonded units of the protein myosin. Myosin is the protein that causes muscles to contract.

- *Thin filaments*. Thin filaments are made of three proteins:

1. *Actin*: Actin forms a helical structure that makes up the bulk of the thin filament mass. Actin contains myosin-binding sites that allow myosin to connect to and move actin during muscle contraction.

2. *Tropomyosin*: Tropomyosin is a long protein fiber that wraps around actin and covers the myosin binding sites on actin.

3. *Troponin*: Bound very tightly to tropomyosin, troponin moves tropomyosin away from myosin binding sites during muscle contraction.

Muscular System Physiology

Function of Muscle Tissue

The main function of the muscular system is movement. Muscles are the only tissue in the body that has the ability to contract and therefore move the other parts of the body.

Related to the function of movement is the muscular system's second function: the maintenance of posture and body position. Muscles often contract to hold the body still or in a particular position rather than to cause movement. The muscles responsible for the body's posture have the greatest endurance of all muscles in the body—they hold up the body throughout the day without becoming tired.

Another function related to movement is the movement of substances inside the body. The cardiac and visceral muscles are primarily responsible for transporting substances like blood or food from one part of the body to another.

The final function of muscle tissue is the generation of body heat. As a result of the high metabolic rate of contracting muscle, our muscular system produces a great deal of waste heat. Many small muscle contractions within the body produce our natural body heat. When we exert ourselves more than normal, the extra muscle contractions lead to a rise in body temperature and eventually to sweating.

Skeletal Muscles as Levers

Skeletal muscles work together with bones and joints to form lever systems. The muscle acts as the effort force; the joint acts as the fulcrum; the bone that the muscle moves acts as the lever; and the object being moved acts as the load.

There are three classes of levers, but the vast majority of the levers in the body are third class levers. A third class lever is a system in which the fulcrum is at the end of the lever and the effort is between the fulcrum and the load at the other end of the lever. The third class levers in the body serve to increase the distance moved by the load compared to the distance that the muscle contracts.

The tradeoff for this increase in distance is that the force required to move the load must be greater than the mass of the load. For example, the biceps brachia of the arm pulls on the radius of the forearm, causing flexion at the elbow joint in a third class lever system. A very slight change in the length of the biceps causes a much larger movement of the forearm and hand, but the force applied by the biceps must be higher than the load moved by the muscle.

Motor Units

Nerve cells called motor neurons control the skeletal muscles. Each motor neuron controls several muscle cells in a group known as a motor unit. When a motor neuron receives a signal from the brain, it stimulates all of the muscles cells in its motor unit at the same time.

The size of motor units varies throughout the body, depending on the function of a muscle. Muscles that perform fine movements—like those of the eyes or fingers—have very few muscle fibers in each motor unit to improve the precision of the brain's control over these structures. Muscles that need a lot of strength to perform their function—like leg or arm muscles—have many muscle cells in each motor unit. One of the ways that the body can control the strength of each muscle is by determining how many motor units to activate for a given function. This explains why the same muscles that are used to pick up a pencil are also used to pick up a bowling ball.

Contraction Cycle

Muscles contract when stimulated by signals from their motor neurons. Motor neurons contact muscle cells at a point called the Neuromuscular Junction (NMJ). Motor neurons release neurotransmitter chemicals at the NMJ that bond to a special part of the sarcolemma known as the motor end plate. The motor end plate contains many ion channels that open in response to neurotransmitters and allow positive ions to enter the muscle fiber. The positive ions form an electrochemical gradient to form inside of the cell, which spreads throughout the sarcolemma and the T-tubules by opening even more ion channels.

When the positive ions reach the sarcoplasmic reticulum, Ca2+ ions are released and allowed to flow into the myofibrils. Ca2+ ions bind to troponin, which causes the troponin molecule to change shape and move nearby molecules of tropomyosin. Tropomyosin is moved away from myosin binding sites on actin molecules, allowing actin and myosin to bind together.

ATP molecules power myosin proteins in the thick filaments to bend and pull on actin molecules in the thin filaments. Myosin proteins act like oars on a boat, pulling the thin filaments closer to the center of a sarcomere. As the thin filaments are pulled together, the sarcomere shortens and contracts. Myofibrils of muscle fibers are made of many sarcomeres in a row, so that when all of the sarcomeres contract, the muscle cells shortens with a great force relative to its size.

Muscles continue contraction as long as they are stimulated by a neurotransmitter. When a motor neuron stops the release of the neurotransmitter, the process of contraction reverses itself. Calcium returns to the sarcoplasmic reticulum; troponin and tropomyosin return to their resting positions; and actin and myosin are prevented from binding. Sarcomeres return to their elongated resting state once the force of myosin pulling on actin has stopped.

Types of Muscle Contraction

The strength of a muscle's contraction can be controlled by two factors: the number of motor units involved in contraction and the amount of stimulus from the nervous system. A single nerve impulse of a motor neuron will cause a motor unit to contract briefly before relaxing. This small contraction is known as a twitch contraction. If the motor neuron provides several signals within a short period of time, the strength and duration of the muscle contraction increases. This phenomenon is known as temporal summation. If the motor neuron provides many nerve impulses in rapid succession, the muscle may enter the state of tetanus, or complete and lasting contraction. A muscle will remain in tetanus until the nerve signal rate slows or until the muscle becomes too fatigued to maintain the tetanus.

Not all muscle contractions produce movement. Isometric contractions are light contractions that increase the tension in the muscle without exerting enough force to move a body part. When people tense their bodies due to stress, they are performing an isometric contraction. Holding an object still and maintaining posture are also the result of isometric contractions. A contraction that does produce movement is an isotonic contraction. Isotonic contractions are required to develop muscle mass through weight lifting.

Muscle tone is a natural condition in which a skeletal muscle stays partially contracted at all times. Muscle tone provides a slight tension on the muscle to prevent damage to the muscle and joints from sudden movements, and also helps to maintain the body's posture. All muscles maintain some amount of muscle tone at all times, unless the muscle has been disconnected from the central nervous system due to nerve damage.

Functional Types of Skeletal Muscle Fibers

Skeletal muscle fibers can be divided into two types based on how they produce and use energy: Type I and Type II.

- Type I fibers are very slow and deliberate in their contractions. They are very resistant to fatigue because they use aerobic respiration to produce energy from sugar. We find Type I fibers in muscles throughout the body for stamina and posture. Near the spine and neck regions, very high concentrations of Type I fibers hold the body up throughout the day.

- Type II fibers are broken down into two subgroups: Type II A and Type II B.

  o Type II A fibers are faster and stronger than Type I fibers, but do not have as much endurance. Type II A fibers are found throughout the body, but especially in the legs where they work to support your body throughout a long day of walking and standing.

  o Type II B fibers are even faster and stronger than Type II A, but have even less endurance. Type II B fibers are also much lighter in color than Type I and Type II A due to their lack of myoglobin, an oxygen-storing pigment. We find Type II B fibers throughout the body, but particularly in the upper body where they give speed and strength to the arms and chest at the expense of stamina.

Muscle Metabolism and Fatigue

Muscles get their energy from different sources depending on the situation that the muscle is working in. Muscles use aerobic respiration when we call on them to produce a low to moderate level of force. Aerobic respiration requires oxygen to produce about 36-38 ATP molecules from a molecule of glucose. Aerobic respiration is very efficient, and can continue as long as a muscle receives adequate amounts of oxygen and glucose to keep contracting. When we use muscles to produce a high level of force, they become so tightly contracted that oxygen carrying blood cannot enter the muscle. This condition causes the muscle to create energy using lactic acid fermentation, a form of anaerobic respiration. Anaerobic respiration is much less efficient than aerobic respiration—only 2 ATP are produced for each molecule of glucose. Muscles quickly tire as they burn through their energy reserves under anaerobic respiration.

To keep muscles working for a longer period of time, muscle fibers contain several important energy molecules. Myoglobin, a red pigment found in muscles, contains iron and stores oxygen in a manner similar to hemoglobin in the blood. The oxygen from myoglobin allows muscles to continue aerobic respiration in the absence of oxygen. Another chemical that helps to keep muscles working is creatine phosphate. Muscles use energy in the form of ATP, converting ATP to ADP to release its energy. Creatine phosphate donates its phosphate group to ADP to turn it back into ATP in order to provide extra energy to the muscle. Finally, muscle fibers contain energy-storing glycogen, a large macromolecule made of many linked glucoses. Active muscles break glucoses off of glycogen molecules to provide an internal fuel supply.

When muscles run out of energy during either aerobic or anaerobic respiration, the muscle quickly tires and loses its ability to contract. This condition is known as muscle fatigue. A fatigued muscle contains very little or no oxygen, glucose or ATP, but instead has many waste products from respiration, like lactic acid and ADP. The body must take in extra oxygen after exertion to replace the oxygen that was stored in myoglobin in the muscle fiber as well as to power the aerobic respiration that will rebuild the energy supplies inside of the cell. Oxygen debt (or recovery oxygen uptake) is the name for the extra oxygen that the body must take in to restore the muscle cells to their resting state. This explains why you feel out of breath for a few minutes after a strenuous activity—your body is trying to restore itself to its normal state.

Cardiovascular System

The cardiovascular system consists of the heart, blood vessels, and the approximately 5 liters of blood that the blood vessels transport. Responsible for transporting oxygen, nutrients, hormones, and cellular waste products throughout the body, the cardiovascular system is powered by the body's hardest-working organ — the heart, which is only about the size of a closed fist. Even at rest, the average heart easily pumps over 5 liters of blood throughout the body every minute.

Cardiovascular System Anatomy

The Heart

The heart is a muscular pumping organ located medial to the lungs along the body's midline in the thoracic region. The bottom tip of the heart, known as its apex, is turned to the left, so that about 2/3 of the heart is located on the body's left side with the other 1/3 on right. The top of the heart, known as the heart's base, connects to the great blood vessels of the body: the aorta, vena cava, pulmonary trunk, and pulmonary veins.

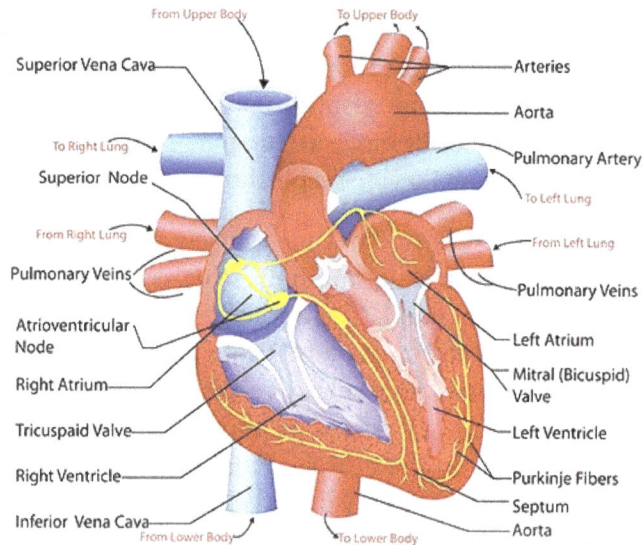

Labels (clockwise): From Upper Body, To Upper Body, Superior Vena Cava, Arteries, Aorta, To Right Lung, Superior Node, Pulmonary Artery, To Left Lung, From Right Lung, From Left Lung, Pulmonary Veins, Pulmonary Veins, Atrioventricular Node, Left Atrium, Right Atrium, Mitral (Bicuspid) Valve, Tricuspaid Valve, Left Ventricle, Right Ventricle, Purkinje Fibers, Septum, Inferior Vena Cava, Aorta, From Lower Body, To Lower Body

Circulatory Loops

There are 2 primary circulatory loops in the human body: the *pulmonary circulation loop* and the *systemic circulation loop*.

1. Pulmonary circulation transports deoxygenated blood from the right side of the heart to the lungs, where the blood picks up oxygen and returns to the left side of the heart. The pumping chambers of the heart that support the pulmonary circulation loop are the right atrium and right ventricle.

2. Systemic circulation carries highly oxygenated blood from the left side of the heart to all of the tissues of the body (with the exception of the heart and lungs). Systemic circulation removes wastes from body tissues and returns deoxygenated blood to the right side of the heart. The left atrium and left ventricle of the heart are the pumping chambers for the systemic circulation loop.

Blood Vessels

Blood vessels are the body's highways that allow blood to flow quickly and efficiently from the heart to every region of the body and back again. The size of blood vessels corresponds with the amount of blood that passes through the vessel. All blood vessels contain a hollow area called the lumen through which blood is able to flow. Around the lumen is the wall of the vessel, which may be thin in the case of capillaries or very thick in the case of arteries.

All blood vessels are lined with a thin layer of simple squamous epithelium known as the endothelium that keeps blood cells inside of the blood vessels and prevents clots from forming. The endothelium lines the entire circulatory system, all the way to the interior of the heart, where it is called the endocardium.

There are three major types of blood vessels: arteries, capillaries and veins. Blood vessels are often named after either the region of the body through which they carry blood or for nearby structures.

For example, the brachiocephalic artery carries blood into the brachial (arm) and cephalic (head) regions. One of its branches, the subclavian artery, runs under the clavicle; hence the name subclavian. The subclavian artery runs into the axillary region where it becomes known as the axillary artery.

Arteries and Arterioles

Arteries are blood vessels that carry blood away from the heart. Blood carried by arteries is usually highly oxygenated, having just left the lungs on its way to the body's tissues. The pulmonary trunk and arteries of the pulmonary circulation loop provide an exception to this rule — these arteries carry deoxygenated blood from the heart to the lungs to be oxygenated.

Arteries face high levels of blood pressure as they carry blood being pushed from the heart under great force. To withstand this pressure, the walls of the arteries are thicker, more elastic, and more muscular than those of other vessels. The largest arteries of the body contain a high percentage of elastic tissue that allows them to stretch and accommodate the pressure of the heart.

Smaller arteries are more muscular in the structure of their walls. The smooth muscles of the arterial walls of these smaller arteries contract or expand to regulate the flow of blood through their lumen. In this way, the body controls how much blood flows to different parts of the body under varying circumstances. The regulation of blood flow also affects blood pressure, as smaller arteries give blood less area to flow through and therefore increases the pressure of the blood on arterial walls.

Arterioles are narrower arteries that branch off from the ends of arteries and carry blood to capillaries. They face much lower blood pressures than arteries due to their greater number, decreased blood volume, and distance from the direct pressure of the heart. Thus arteriole walls are much thinner than those of arteries. Arterioles, like arteries, are able to use smooth muscle to control their aperture and regulate blood flow and blood pressure.

Capillaries

Capillaries are the smallest and thinnest of the blood vessels in the body and also the most common. They can be found running throughout almost every tissue of the body and border the edges of the body's avascular tissues. Capillaries connect to arterioles on one end and venules on the other.

Capillaries carry blood very close to the cells of the tissues of the body in order to exchange gases, nutrients, and waste products. The walls of capillaries consist of only a thin layer of endothelium so that there is the minimum amount of structure possible between the blood and the tissues. The endothelium acts as a filter to keep blood cells inside of the vessels while allowing liquids, dissolved gases, and other chemicals to diffuse along their concentration gradients into or out of tissues.

Precapillary sphincters are bands of smooth muscle found at the arteriole ends of capillaries. These sphincters regulate blood flow into the capillaries. Since there is a limited supply of blood, and not all tissues have the same energy and oxygen requirements, the precapillary sphincters reduce blood flow to inactive tissues and allow free flow into active tissues.

Veins and Venules

Veins are the large return vessels of the body and act as the blood return counterparts of arteries. Because the arteries, arterioles, and capillaries absorb most of the force of the heart's contractions, veins and venules are subjected to very low blood pressures. This lack of pressure allows the walls of veins to be much thinner, less elastic, and less muscular than the walls of arteries.

Veins rely on gravity, inertia, and the force of skeletal muscle contractions to help push blood back to the heart. To facilitate the movement of blood, some veins contain many one-way valves that prevent blood from flowing away from the heart. As skeletal muscles in the body contract, they squeeze nearby veins and push blood through valves closer to the heart.

When the muscle relaxes, the valve traps the blood until another contraction pushes the blood closer to the heart. Venules are similar to arterioles as they are small vessels that connect capillaries, but unlike arterioles, venules connect to veins instead of arteries. Venules pick up blood from many capillaries and deposit it into larger veins for transport back to the heart.

Coronary Circulation

The heart has its own set of blood vessels that provide the myocardium with the oxygen and nutrients necessary to pump blood throughout the body. The left and right coronary arteries branch off from the aorta and provide blood to the left and right sides of the heart. The coronary sinus is a vein on the posterior side of the heart that returns deoxygenated blood from the myocardium to the vena cava.

Hepatic Portal Circulation

The veins of the stomach and intestines perform a unique function: instead of carrying blood directly back to the heart, they carry blood to the liver through the hepatic portal vein. Blood leaving the digestive organs is rich in nutrients and other chemicals absorbed from food. The liver removes toxins, stores sugars, and processes the products of digestion before they reach the other body tissues. Blood from the liver then returns to the heart through the inferior vena cava.

Blood

The average human body contains about 4 to 5 liters of blood. As a liquid connective tissue, it transports many substances through the body and helps to maintain homeostasis of nutrients, wastes, and gases. Blood is made up of red blood cells, white blood cells, platelets, and liquid plasma.

Red Blood Cells

Red blood cells, also known as erythrocytes, are by far the most common type of blood cell and make up about 45% of blood volume. Erythrocytes are produced inside of red bone marrow from stem cells at the astonishing rate of about 2 million cells every second. The shape of erythrocytes is biconcave—disks with a concave curve on both sides of the disk so that the center of an erythrocyte is its thinnest part. The unique shape of erythrocytes gives these cells a high surface area to volume ratio and allows them to fold to fit into thin capillaries. Immature erythrocytes have a nucleus that is ejected from the cell when it reaches maturity to provide it with its unique shape and flexibility. The lack of a nucleus means that red blood cells contain no DNA and are not able to repair themselves once damaged.

Erythrocytes transport oxygen in the blood through the red pigment hemoglobin. Hemoglobin contains iron and proteins joined to greatly increase the oxygen carrying capacity of erythrocytes. The high surface area to volume ratio of erythrocytes allows oxygen to be easily transferred into the cell in the lungs and out of the cell in the capillaries of the systemic tissues.

White Blood Cells

White blood cells, also known as leukocytes, make up a very small percentage of the total number of cells in the bloodstream, but have important functions in the body's immune system. There are two major classes of white blood cells: granular leukocytes and agranular leukocytes.

1. Granular Leukocytes: The three types of granular leukocytes are neutrophils, eosinophils, and basophils. Each type of granular leukocyte is classified by the presence of chemical-filled vesicles in their cytoplasm that give them their function. Neutrophils contain digestive enzymes that neutralize bacteria that invade the body. Eosinophils contain digestive enzymes specialized for digesting viruses that have been bound to by antibodies in the blood. Basophils release histamine to intensify allergic reactions and help protect the body from parasites.

2. Agranular Leukocytes: The two major classes of agranular leukocytes are lymphocytes and monocytes. Lymphocytes include T cells and natural killer cells that fight off viral infections and B cells that produce antibodies against infections by pathogens. Monocytes develop into cells called macrophages that engulf and ingest pathogens and the dead cells from wounds or infections.

Platelets

Also known as thrombocytes, platelets are small cell fragments responsible for the clotting of blood and the formation of scabs. Platelets form in the red bone marrow from large megakaryocyte cells that periodically rupture and release thousands of pieces of membrane that become the platelets. Platelets do not contain a nucleus and only survive in the body for up to a week before macrophages capture and digest them.

Plasma

Plasma is the non-cellular or liquid portion of the blood that makes up about 55% of the blood's volume. Plasma is a mixture of water, proteins, and dissolved substances. Around 90% of plasma is made of water, although the exact percentage varies depending upon the hydration levels of the individual. The proteins within plasma include antibodies and albumins. Antibodies are part of the immune system and bind to antigens on the surface of pathogens that infect the body. Albumins help maintain the body's osmotic balance by providing an isotonic solution for the cells of the body. Many different substances can be found dissolved in the plasma, including glucose, oxygen, carbon dioxide, electrolytes, nutrients, and cellular waste products. The plasma functions as a transportation medium for these substances as they move throughout the body.

Lungs

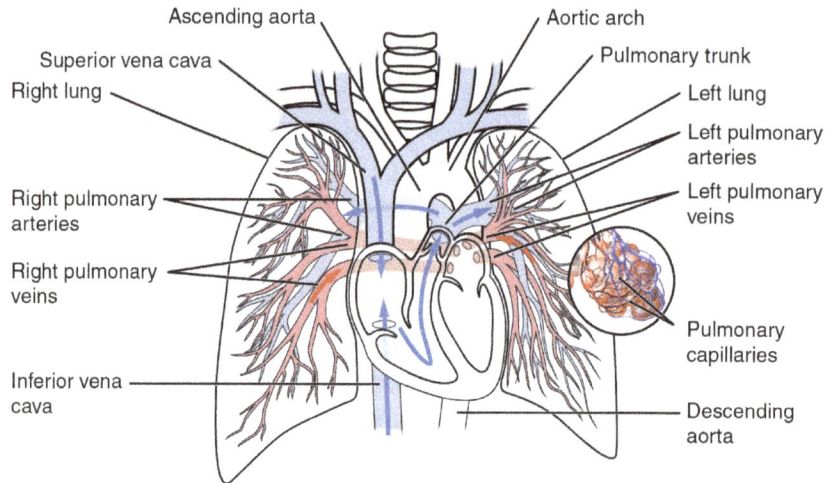

Ascending aorta
Aortic arch
Superior vena cava
Pulmonary trunk
Right lung
Left lung
Left pulmonary arteries
Right pulmonary arteries
Left pulmonary veins
Right pulmonary veins
Pulmonary capillaries
Inferior vena cava
Descending aorta

The pulmonary circulation as it passes from the heart. Showing both the pulmonary and bronchial arteries

The circulatory system of the lungs is the portion of the cardiovascular system in which oxygen-depleted blood is pumped away from the heart, via the pulmonary artery, to the lungs and returned, oxygenated, to the heart via the pulmonary vein.

Oxygen deprived blood from the superior and inferior vena cava enters the right atrium of the heart and flows through the tricuspid valve (right atrioventricular valve) into the right ventricle, from which it is then pumped through the pulmonary semilunar valve into the pulmonary artery to the lungs. Gas exchange occurs in the lungs, whereby CO_2 is released from the blood, and oxygen is absorbed. The pulmonary vein returns the now oxygen-rich blood to the left atrium.

A separate system known as the bronchial circulation supplies blood to the tissue of the larger airways of the lung.

Brain

The brain has a dual blood supply that comes from arteries at its front and back. These are called the "anterior" and "posterior" circulation respectively. The anterior circulation arises from the internal carotid arteries and supplies the front of the brain. The posterior circulation arises from the vertebral arteries, and supplies the back of the brain and brainstem. The circulation from the front and the back join together (anastomise) at the Circle of Willis.

Kidneys

The renal circulation receives around 20% of the cardiac output. It branches from the abdominal aorta and returns blood to the ascending vena cava. It is the blood supply to the kidneys, and contains many specialized blood vessels.

Lymphatic System

The lymphatic system is part of the circulatory system. It is a network of lymphatic vessels and lymph capillaries, lymph nodes and organs, and lymphatic tissues and circulating lymph. One of its major functions is to carry the lymph, draining and returning interstitial fluid back towards the heart for return to the cardiovascular system, by emptying into the lymphatic ducts. Its other main function is in the adaptive immune system.

Cardiovascular System Physiology

Functions of the Cardiovascular System

The cardiovascular system has three major functions: transportation of materials, protection from pathogens, and regulation of the body's homeostasis.

- *Transportation*: The cardiovascular system transports blood to almost all of the body's tissues. The blood delivers essential nutrients and oxygen and removes wastes and carbon dioxide to be processed or removed from the body. Hormones are transported throughout the body via the blood's liquid plasma.

- *Protection*: The cardiovascular system protects the body through its white blood cells. White blood cells clean up cellular debris and fight pathogens that have entered the body. Platelets and red blood cells form scabs to seal wounds and prevent pathogens from entering the body and liquids from leaking out. Blood also carries antibodies that provide specific immunity to pathogens that the body has previously been exposed to or has been vaccinated against.

- *Regulation*: The cardiovascular system is instrumental in the body's ability to maintain homeostatic control of several internal conditions. Blood vessels help maintain a stable body temperature by controlling the blood flow to the surface of the skin. Blood vessels near the skin's surface open during times of overheating to allow hot blood to dump its heat into the body's surroundings. In the case of hypothermia, these blood vessels constrict to keep blood flowing only to vital organs in the body's core. Blood also helps balance the body's pH due to the presence of bicarbonate ions, which act as a buffer solution. Finally, the albumins in blood plasma help to balance the osmotic concentration of the body's cells by maintaining an isotonic environment.

Many serious conditions and diseases can cause our cardiovascular system to stop working properly. Quite often, we don't do enough about them proactively, resulting in emergencies. Browse our content to learn more about cardiovascular health. Also, explore how DNA health testing can allow you to begin important conversations with your doctor about genetic risks for disorders involving clotting, hemophilia, hemochromatosis (a common hereditary disorder causing iron to accumulate in the heart) and glucose-6-phosphate dehydrogenase (which affects about 1 in 10 African American men).

The Circulatory Pump

The heart is a four-chambered "double pump," where each side (left and right) operates as a separate pump. The left and right sides of the heart are separated by a muscular wall of tissue known as the septum of the heart. The right side of the heart receives deoxygenated blood from the systemic veins and pumps it to the lungs for oxygenation. The left side of the heart receives oxygenated blood from the lungs and pumps it through the systemic arteries to the tissues of the body. Each heartbeat results in the simultaneous pumping of both sides of the heart, making the heart a very efficient pump.

Regulation of Blood Pressure

Several functions of the cardiovascular system can control blood pressure. Certain hormones along with autonomic nerve signals from the brain affect the rate and strength of heart contractions. Greater contractile force and heart rate lead to an increase in blood pressure. Blood vessels can also affect blood pressure. Vasoconstriction decreases the diameter of an artery by contracting the smooth muscle in the arterial wall. The sympathetic (fight or flight) division of the autonomic nervous system causes vasoconstriction, which leads to increases in blood pressure and decreases in blood flow in the constricted region. Vasodilation is the expansion of an artery as the smooth muscle in the arterial wall relaxes after the fight-or-flight response wears off or under the effect of certain hormones or chemicals in the blood. The volume of blood in the body also affects blood pressure. A higher volume of blood in the body raises blood pressure by increasing the amount of blood pumped by each heartbeat. Thicker, more viscous blood from clotting disorders can also raise blood pressure.

Hemostasis

Hemostasis, or the clotting of blood and formation of scabs, is managed by the platelets of the blood. Platelets normally remain inactive in the blood until they reach damaged tissue or leak out of the blood vessels through a wound. Once active, platelets change into a spiny ball shape and become very sticky in order to latch on to damaged tissues. Platelets next release chemical clotting factors and begin to produce the protein fibrin to act as structure for the blood clot. Platelets also begin sticking together to form a platelet plug. The platelet plug will serve as a temporary seal to keep blood in the vessel and foreign material out of the vessel until the cells of the blood vessel can repair the damage to the vessel wall.

Digestive System

The digestive system is made up of the gastrointestinal tract—also called the GI tract or digestive tract—and the liver, pancreas, and gallbladder. The GI tract is a series of hollow organs joined in a long, twisting tube from the mouth to the anus. The hollow organs that make up the GI tract are the mouth, esophagus, stomach, small intestine, large intestine, and anus. The liver, pancreas, and gallbladder are the solid organs of the digestive system.

The small intestine has three parts. The first part is called the duodenum. The jejunum is in the middle and the ileum is at the end. The large intestine includes the appendix, cecum, colon, and rectum. The appendix is a finger-shaped pouch attached to the cecum. The cecum is the first part of the large intestine. The colon is next. The rectum is the end of the large intestine.

Bacteria in your GI tract, also called gut flora or microbiome, help with digestion. Parts of your nervous and circulatory systems also help. Working together, nerves, hormones, bacteria, blood, and the organs of your digestive system digest the foods and liquids you eat or drink each day.

The Digestive System

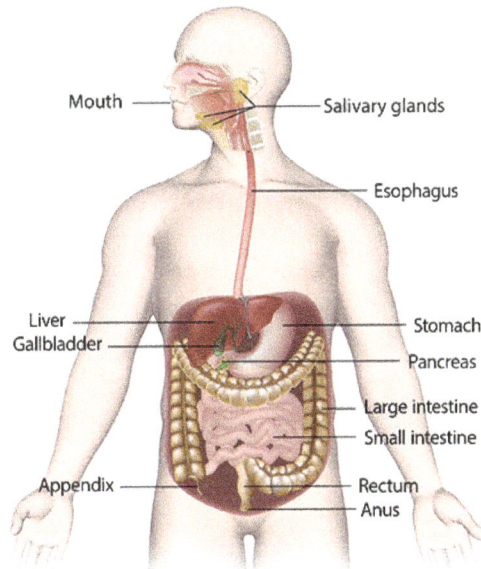

The digestive system

Importance of Digestion

Digestion is important because your body needs nutrients from food and drink to work properly and stay healthy. Proteins, fats, carbohydrates, vitamins , minerals , and water are nutrients. Your digestive system breaks nutrients into parts small enough for your body to absorb and use for energy, growth, and cell repair.

- Proteins break into amino acids

- Fats break into fatty acids and glycerol

- Carbohydrates break into simple sugars

Your digestive system breaks nutrients into parts that are small enough for your body to absorb

Components

There are several organs and other components involved in the digestion of food. The organs known as the accessory digestive glands are the liver, gall bladder and pancreas. Other components include the mouth, salivary glands, tongue, teeth and epiglottis.

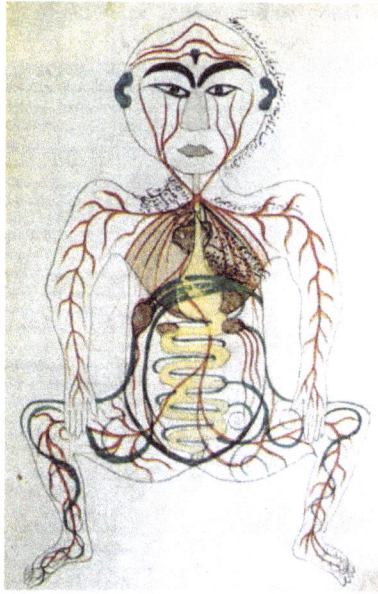

Historical depiction of the digestive system

The largest structure of the digestive system is the gastrointestinal tract (GI tract). This starts at the mouth and ends at the anus, covering a distance of about nine (9) metres.

The largest part of the GI tract is the colon or large intestine. Water is absorbed here and the remaining waste matter is stored prior to defecation.

Most of the digestion of food takes place in the small intestine.

A major digestive organ is the stomach. Within its mucosa are millions of embedded gastric glands. Their secretions are vital to the functioning of the organ.

There are many specialised cells of the GI tract. These include the various cells of the gastric glands, taste cells, pancreatic duct cells, enterocytes and microfold cells.

Some parts of the digestive system are also part of the excretory system, including the large intestine.

Mouth

The mouth is the first part of the gastrointestinal tract and is equipped with several structures that begin the first processes of digestion. These include salivary glands, teeth and the tongue. The mouth consists of two regions; the vestibule and the oral cavity proper. The vestibule is the area between the teeth, lips and cheeks, and the rest is the oral cavity proper. Most of the oral cavity is lined with oral mucosa, a mucous membrane that produces a lubricating mucus, of which only a small amount is needed. Mucous membranes vary in structure in the different regions of the body but they all produce a lubricating mucus, which is either secreted by surface cells or more usually by underlying glands. The mucous membrane in the mouth continues as the thin mucosa which lines the bases of the teeth. The main component of mucus is a glycoprotein called mucin and the type secreted varies according to the region involved. Mucin is viscous, clear, and clinging. Underlying the mucous membrane in the mouth is a thin layer of smooth muscle tissue and the loose connection to the membrane gives it its great elasticity. It covers the cheeks, inner surfaces of the lips, and floor of the mouth.

The roof of the mouth is termed the palate and it separates the oral cavity from the nasal cavity. The palate is hard at the front of the mouth since the overlying mucosa is covering a plate of bone; it is softer and more pliable at the back being made of muscle and connective tissue, and it can move to swallow food and liquids. The soft palate ends at the uvula. The surface of the hard palate allows for the pressure needed in eating food, to leave the nasal passage clear. The lips are the mouth's front boundary and the fauces (the passageway between the tonsils, also called the throat), mark its posterior boundary.

At either side of the soft palate are the palatoglossus muscles which also reach into regions of the tongue. These muscles raise the back of the tongue and also close both sides of the fauces to enable food to be swallowed. Mucus helps in the mastication of food in its ability to soften and collect the food in the formation of the bolus.

Salivary Glands

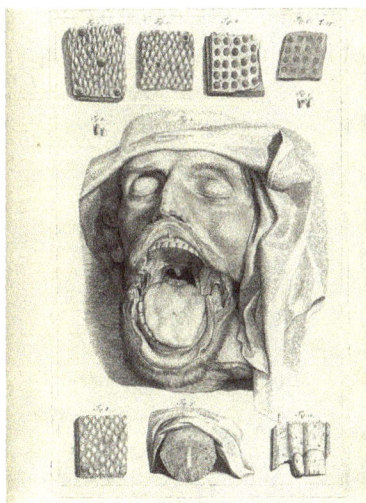

Oral cavity

There are three pairs of main salivary glands and between 800 and 1,000 minor salivary glands, all of which mainly serve the digestive process, and also play an important role in the maintenance of dental health and general mouth lubrication, without which speech would be impossible. The main glands are all exocrine glands, secreting via ducts. All of these glands terminate in the mouth. The largest of these are the parotid glands—their secretion is mainly serous. The next pair are underneath the jaw, the submandibular glands, these produce both serous fluid and mucus. The serous fluid is produced by serous glands in these salivary glands which also produce lingual lipase. They produce about 70% of the oral cavity saliva. The third pair are the sublingual glands located underneath the tongue and their secretion is mainly mucous with a small percentage of saliva.

Within the oral mucosa (a mucous membrane) lining the mouth and also on the tongue and palates and mouth floor, are the minor salivary glands; their secretions are mainly mucous and are innervated by the facial nerve (the seventh cranial nerve). The glands also secrete amylase a first stage in the breakdown of food acting on the carbohydrate in the food to transform the starch content into maltose. There are other glands on the surface of the tongue that encircle taste buds on the back part of the tongue and these also produce lingual lipase. Lipase is a digestive enzyme that catalyses the hydrolysis of lipids (fats). These glands are termed Von Ebner's glands which have also been shown to have another function in the secretion of histatins which offer an early defense (outside of the immune system) against microbes in food, when it makes contact with these glands on the tongue tissue. Sensory information can stimulate the secretion of saliva providing the necessary fluid for the tongue to work with and also to ease swallowing of the food.

Saliva

Saliva moistens and softens food, and along with the chewing action of the teeth, transforms the food into a smooth bolus. The bolus is further helped by the lubrication provided by the saliva in its passage from the mouth into the esophagus. Also of importance is the presence in saliva of the digestive enzymes amylase and lipase. Amylase starts to work on the starch in carbohydrates, breaking it down into the simple sugars of maltose and dextrose that can be further broken down in the small intestine. Saliva in the mouth can account for 30% of this initial starch digestion. Lipase starts to work on breaking down fats. Lipase is further produced in the pancreas where it is released to continue this digestion of fats. The presence of salivary lipase is of prime importance in young babies whose pancreatic lipase has yet to be developed.

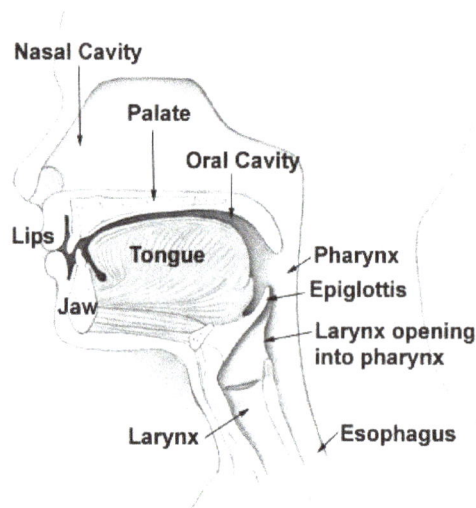

As well as its role in supplying digestive enzymes, saliva has a cleansing action for the teeth and mouth. It also has an immunological role in supplying antibodies to the system, such as immuno-globulin A. This is seen to be key in preventing infections of the salivary glands, importantly that of parotitis.

Saliva also contains a glycoprotein called haptocorrin which is a binding protein to vitamin B_{12}. It binds with the vitamin in order to carry it safely through the acidic content of the stomach. When it reaches the duodenum, pancreatic enzymes break down the glycoprotein and free the vitamin which then binds with intrinsic factor.

Tongue

Food enters the mouth where the first stage in the digestive process takes place, with the action of the tongue and the secretion of saliva. The tongue is a fleshy and muscular sensory organ, and the very first sensory information is received via the taste buds in the papillae on its surface. If the taste is agreeable, the tongue will go into action, manipulating the food in the mouth which stimulates the secretion of saliva from the salivary glands. The liquid quality of the saliva will help in the softening of the food and its enzyme content will start to break down the food whilst it is still in the mouth. The first part of the food to be broken down is the starch of carbohydrates (by the enzyme amylase in the saliva).

The tongue is attached to the floor of the mouth by a ligamentous band called the frenum and this gives it great mobility for the manipulation of food (and speech); the range of manipulation is optimally controlled by the action of several muscles and limited in its external range by the stretch of the frenum. The tongue's two sets of muscles, are four intrinsic muscles that originate in the tongue and are involved with its shaping, and four extrinsic muscles originating in bone that are involved with its movement.

Taste

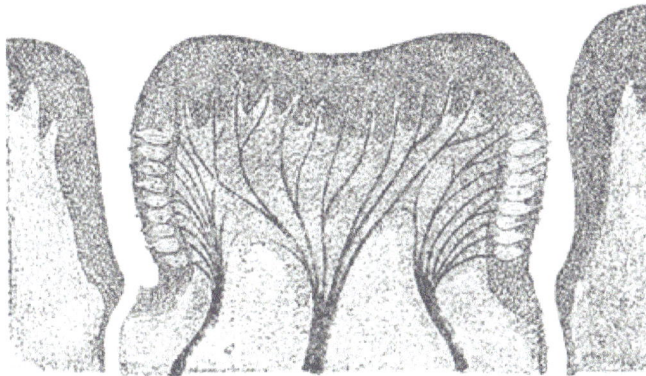

Cross section of circumvallate papilla showing arrangement of nerves and taste buds

Taste is a form of chemoreception that takes place in the specialised taste receptors, contained in structures called taste buds in the mouth. Taste buds are mainly on the upper surface (dorsum) of the tongue. The function of taste perception is vital to help prevent harmful or rotten foods from being consumed. There are also taste buds on the epiglottis and upper part of the esophagus. The taste buds are innervated by a branch of the facial nerve the chorda tympani, and the glossopharyngeal nerve. Taste messages are sent via these cranial nerves to the brain. The brain can distinguish between the chemical qualities of the food. The five basic tastes are referred to as those of saltiness, sourness, bitterness, sweetness, and umami. The detection of saltiness and sourness enables the control of salt and acid balance. The detection of bitterness warns of poisons—many of a plant's defences are of poisonous compounds that are bitter. Sweetness guides to those foods that will supply energy; the initial breakdown of the energy-giving carbohydrates by salivary amylase creates the taste of sweetness since simple sugars are the first result. The taste of umami is thought to signal protein-rich food. Sour tastes are acidic which is often found in bad food. The brain has to decide very quickly whether the food should be eaten or not. It was the findings in 1991, describing the first olfactory receptors that helped to prompt the research into taste. The olfactory receptors are located on cell surfaces in the nose which bind to chemicals enabling the detection of smells. It is assumed that signals from taste receptors work together with those from the nose, to form an idea of complex food flavours.

Teeth

Teeth are complex structures made of materials specific to them. They are made of a bone-like material called dentin, which is covered by the hardest tissue in the body—enamel. Teeth have different shapes to deal with different aspects of mastication employed in tearing and chewing pieces of food into smaller and smaller pieces. This results in a much larger surface area for the action of digestive enzymes. The teeth are named after their particular roles in the process of mastication—incisors are used for cutting or biting off pieces of food; canines, are used for tearing, premolars and molars are used for chewing and grinding. Mastication of the food with the help of saliva and mucus results in the formation of a soft bolus which can then be swallowed to make its way down the upper gastrointestinal tract to the stomach. The digestive enzymes in saliva also help in keeping the teeth clean by breaking down any lodged food particles.

Epiglottis

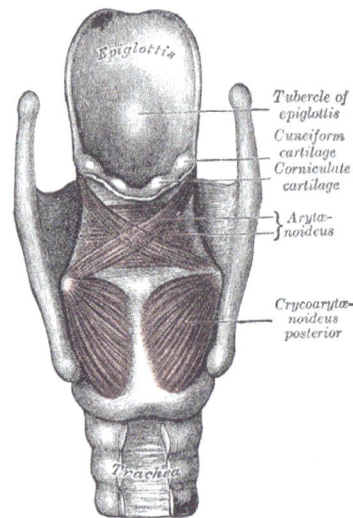

The epiglottis is a flap of elastic cartilage attached to the entrance of the larynx. It is covered with a mucous membrane and there are taste buds on its lingual surface which faces into the mouth. Its laryngeal surface faces into the larynx. The epiglottis functions to guard the entrance of the glottis, the opening between the vocal folds. It is normally pointed upward during breathing with its underside functioning as part of the pharynx, but during swallowing, the epiglottis folds down to a more horizontal position, with its upper side functioning as part of the pharynx. In this manner it prevents food from going into the trachea and instead directs it to the esophagus, which is behind. During swallowing, the backward motion of the tongue forces the epiglottis over the glottis' opening to prevent any food that is being swallowed from entering the larynx which leads to the lungs; the larynx is also pulled upwards to assist this process. Stimulation of the larynx by ingested matter produces a strong cough reflex in order to protect the lungs.

Pharynx

The pharynx is a part of the conducting zone of the respiratory system and also a part of the digestive system. It is the part of the throat immediately behind the nasal cavity at the back of the mouth and above the esophagus and larynx. The pharynx is made up of three parts. The lower two parts—the oropharynx and the laryngopharynx are involved in the digestive system. The laryngo-pharynx connects to the esophagus and it serves as a passageway for both air and food. Air enters the larynx anteriorly but anything swallowed has priority and the passage of air is temporarily blocked. The pharynx is innervated by the pharyngeal plexus of the vagus nerve. Muscles in the pharynx push the food into the esophagus. The pharynx joins the esophagus at the oesophageal inlet which is located behind the cricoid cartilage.

Esophagus

The esophagus, commonly known as the gullet, is an organ which consists of a muscular tube through which food passes from the pharynx to the stomach. The esophagus is continuous with the laryngeal part of the pharynx. It passes through the posterior mediastinum in the thorax and enters the stomach through a hole in the thoracic diaphragm—the esophageal hiatus, at the level of the tenth thoracic vertebra (T10). Its length averages 25 cm, varying with height. It is divided into cervical, thoracic and abdominal parts. The pharynx joins the esophagus at the esophageal inlet which is behind the cricoid cartilage.

At rest the esophagus is closed at both ends, by the upper and lower esophageal sphincters. The opening of the upper sphincter is triggered by the swallowing reflex so that food is allowed through. The sphincter also serves to prevent back flow from the esophagus into the pharynx. The esophagus has a mucous membrane and the epithelium which has a protective function is continuously replaced due to the volume of food that passes inside the esophagus. During swallowing, food passes from the mouth through the pharynx into the esophagus. The epiglottis folds down to a more horizontal position so as to prevent food from going into the trachea, instead directing it to the esophagus.

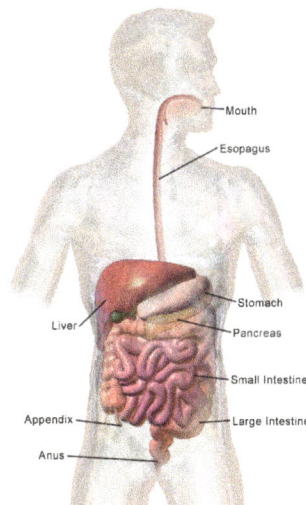

Digestive system in an adult

Once in the esophagus, the bolus travels down to the stomach via rhythmic contraction and re-laxation of muscles known as peristalsis. The lower esophageal sphincter is a muscular sphincter surrounding the lower part of the esophagus. The junction between the esophagus and the stom-ach (the gastroesophageal junction) is controlled by the lower esophageal sphincter, which re-mains constricted at all times other than during swallowing and vomiting to prevent the contents of the stomach from entering the esophagus. As the esophagus does not have the same protection from acid as the stomach, any failure of this sphincter can lead to heartburn. The esophagus has a mucous membrane of epithelium which has a protective function as well as providing a smooth surface for the passage of food. Due to the high volume of food that is passed over time, this membrane is continuously renewed.

Diaphragm

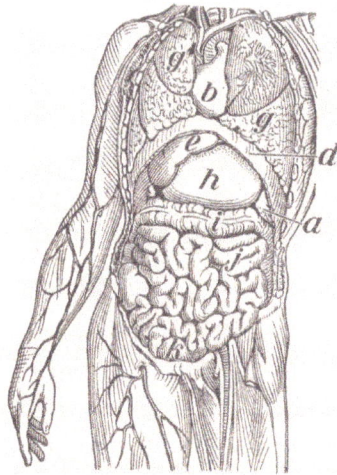

Front view of the viscera. *a*, spleen; *b*, heart; *d*, diaphragm; *e*, liver; *g*, lung; *h*, stomach; *i*, large intestine; *j*, small intestine; *k*, bladdar

The diaphragm is an important part of the body's digestive system. The muscular diaphragm separates the thoracic cavity from the abdominal cavity where most of the digestive organs are located. The suspensory muscle attaches the ascending duodenum to the diaphragm. This muscle is thought to be of help in the digestive system in that its attachment offers a wider angle to the duodenojejunal flexure for the easier passage of digesting material. The diaphragm also attaches to, and anchors the liver at its bare area. The esophagus enters the abdomen through a hole in the diaphragm at the level of T10.

Stomach

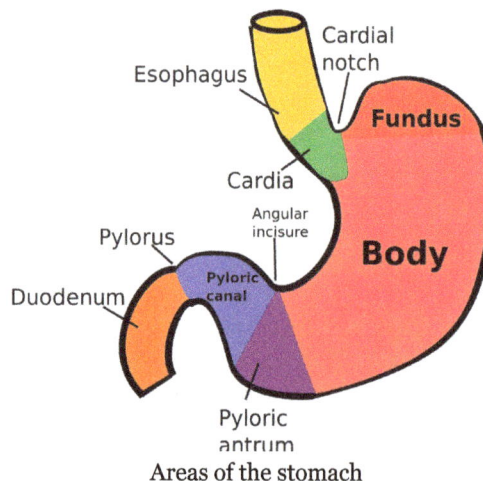

Areas of the stomach

The stomach is a major organ of the gastrointestinal tract and digestive system. It is a consistently J-shaped organ joined to the esophagus at its upper end and to the duodenum at its lower end. Gastric acid (informally *gastric juice*), produced in the stomach plays a vital role in the digestive process, and mainly contains hydrochloric acid and sodium chloride. A peptide hormone, gastrin, produced by G cells in the gastric glands, stimulates the production of gastric juice which activates the digestive enzymes. Pepsinogen is a precursor enzyme (zymogen) produced by the gastric chief cells, and gastric acid activates this to the enzyme pepsin which begins the digestion of proteins. As these two chemicals would damage the stomach wall, mucus is secreted by innumerable gastric glands in the stomach, to provide a slimy protective layer against the damaging effects of the chemicals on the inner layers of the stomach.

At the same time that protein is being digested, mechanical churning occurs through the action of peristalsis, waves of muscular contractions that move along the stomach wall. This allows the mass of food to further mix with the digestive enzymes. Gastric lipase secreted by the chief cells in the fundic glands in the gastric mucosa of the stomach, is an acidic lipase, in contrast with the alkaline pancreatic lipase. This breaks down fats to some degree though is not as efficient as the pancreatic lipase.

The pylorus, the lowest section of the stomach which attaches to the duodenum via the pyloric canal, contains countless glands which secrete digestive enzymes including gastrin. After an hour or two, a thick semi-liquid called chyme is produced. When the pyloric sphincter, or valve opens, chyme enters the duodenum where it mixes further with digestive enzymes from the pancreas, and then passes through the small intestine, where digestion continues. When the chyme is fully digested, it is absorbed into the blood. 95% of absorption of nutrients occurs in the small intestine. Water and minerals are reabsorbed back into the blood in the colon of the large intestine, where the environment is slightly acidic. Some vitamins, such as biotin and vitamin K produced by bacteria in the gut flora of the colon are also absorbed.

The parietal cells in the fundus of the stomach, produce a glycoprotein called intrinsic factor which is essential for the absorption of vitamin B12. Vitamin B12 (cobalamin), is carried to, and through the stomach, bound to a glycoprotein secreted by the salivary glands - transcobalamin I also called haptocorrin, which protects the acid-sensitive vitamin from the acidic stomach contents. Once in the more neutral duodenum, pancreatic enzymes break down the protective glycoprotein. The freed vitamin B12 then binds to intrinsic factor which is then absorbed by the enterocytes in the ileum.

The stomach is a distensible organ and can normally expand to hold about one litre of food. This expansion is enabled by a series of gastric folds in the inner walls of the stomach. The stomach of a newborn baby will only be able to expand to retain about 30 ml.

Spleen

The spleen breaks down both red and white blood cells that are *spent*. This is why it is sometimes known as the 'graveyard of red blood cells'. A product of this *digestion* is the pigment bilirubin, which is sent to the liver and secreted in the bile. Another product is iron, which is used in the formation of new blood cells in the bone marrow. Medicine treats the spleen solely as belonging to the lymphatic system, though it is acknowledged that the full range of its important functions is not yet understood.

Liver

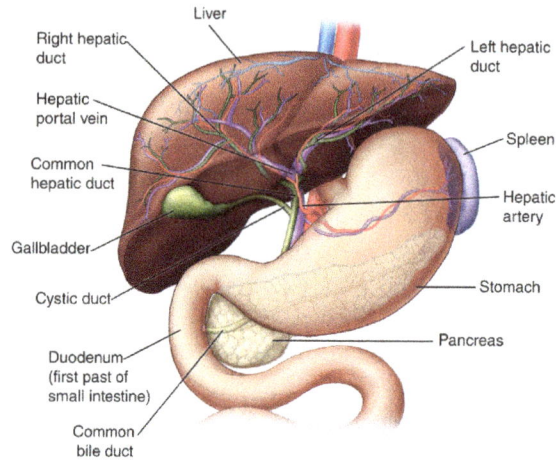

The liver is the second largest organ (after the skin) and is an accessory digestive gland which plays a role in the body's metabolism. The liver has many functions some of which are important to digestion. The liver can detoxify various metabolites; synthesise proteins and produce biochemicals needed for digestion. It regulates the storage of glycogen which it can form from glucose (glycogenesis). The liver can also synthesise glucose from certain amino acids. Its digestive functions are largely involved with the breaking down of carbohydrates. It also maintains protein metabolism in its synthesis and degradation. In lipid metabolism it synthesises cholesterol. Fats are also produced in the process of lipogenesis. The liver synthesises the bulk of lipoproteins. The liver is located in the upper right quadrant of the abdomen and below the diaphragm to which it is attached at one part, This is to the right of the stomach and it overlies the gall bladder. The liver produces bile, an important alkaline compound which aids digestion.

Bile

Bile produced by the liver is made up of water (97%), bile salts, mucus and pigments, 1% fats and inorganic salts. Bilirubin is its major pigment. Bile acts partly as a surfactant which lowers the surface tension between either two liquids or a solid and a liquid and helps to emulsify the fats in the chyme. Food fat is dispersed by the action of bile into smaller units called micelles. The breaking down into micelles creates a much larger surface area for the pancreatic enzyme, lipase to work on. Lipase digests the triglycerides which are broken down into two fatty acids and a monoglyceride. These are then absorbed by villi on the intestinal wall. If fats are not absorbed in this way in the small intestine problems can arise later in the large intestine which is not equipped to absorb fats. Bile also helps in the absorption of vitamin K from the diet. Bile is collected and delivered through the common hepatic duct. This duct joins with the cystic duct to connect in a common bile duct with the gallbladder. Bile is stored in the gallbladder for release when food is discharged into the duodenum and also after a few hours.

Gallbladder

The gallbladder is a hollow part of the biliary tract that sits just beneath the liver, with the gallbladder body resting in a small depression. It is a small organ where the bile produced by the liver is stored, before being released into the small intestine. Bile flows from the liver through the bile ducts and into the gall bladder for storage. The bile is released in response to cholecystokinin (CCK) a peptide hormone released from the duodenum. The production of CCK (by endocrine cells of the duodenum) is stimulated by the presence of fat in the duodenum.

It is divided into three sections, a fundus, body and neck. The neck tapers and connects to the biliary tract via the cystic duct, which then joins the common hepatic duct to form the common bile duct. At this junction is a mucosal fold called *Hartmann's pouch*, where gallstones commonly get stuck. The muscular layer of the body is of smooth muscle tissue that helps the gallbladder contract, so that it can discharge its bile into the bile duct. The gallbladder needs to store bile in a natural, semi-liquid form at all times. Hydrogen ions secreted from the inner lining of the gall-bladder keep the bile acidic enough to prevent hardening. To dilute the bile, water and electrolytes from the digestion system are added. Also, salts attach themselves to cholesterol molecules in the bile to keep them from crystallising. If there is too much cholesterol or bilirubin in the bile, or if the gallbladder doesn't empty properly the systems can fail. This is how gallstones form when a small piece of calcium gets coated with either cholesterol or bilirubin and the bile crystallises and forms a gallstone. The main purpose of the gallbladder is to store and release bile, or *gall*. Bile is released into the small intestine in order to help in the digestion of fats by breaking down larger molecules into smaller ones. After the fat is absorbed, the bile is also absorbed and transported back to the liver for reuse.

Pancreas

The pancreas is a major organ functioning as an accessory digestive gland in the digestive system. It is both an endocrine gland and an exocrine gland. The endocrine part secretes insulin when the blood sugar becomes high; insulin moves glucose from the blood into the muscles and other tissues for use as energy. The endocrine part releases glucagon when the blood sugar is low; glucagon allows stored sugar to be broken down into glucose by the liver in order to re-balance the sugar levels. The pancreas produces and releases important digestive enzymes in the pancreatic juice that it delivers to the duodenum. The pancreas lies below and at the back of the stomach. It connects to the duode-num via the pancreatic duct which it joins near to the bile duct's connection where both the bile and pancreatic juice can act on the chyme that is released from the stomach into the duodenum. Aqueous pancreatic secretions from pancreatic duct cells contain bicarbonate ions which are alkaline and help with the bile to neutralise the acidic chyme that is churned out by the stomach.

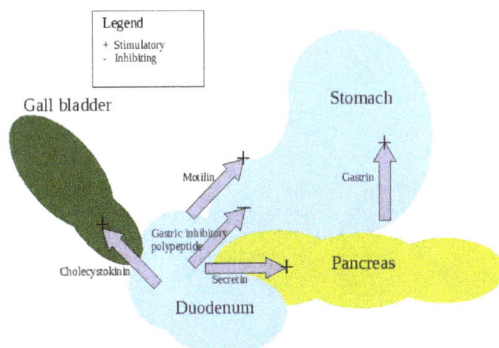

Action of digestive hormones

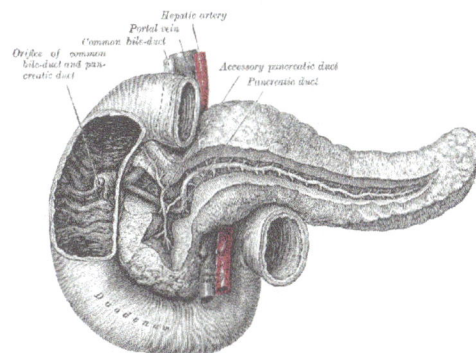

Pancreas, duodenum and bile duct

The pancreas is also the main source of enzymes for the digestion of fats and proteins. Some of these are released in response to the production of CKK in the duodenum. (The enzymes that digest polysaccharides, by contrast, are primarily produced by the walls of the intestines.) The cells are filled with secretory granules containing the precursor digestive enzymes. The major proteases, the pancreatic enzymes which work on proteins, are trypsinogen and chymotrypsinogen. Elastase is also produced. Smaller amounts of lipase and amylase are secreted. The pancreas also secretes phospholipase A2, lysophospholipase, and cholesterol esterase. The precursor zymogens, are inactive variants of the enzymes; which avoids the onset of pancreatitis caused by autodegradation. Once released in the intestine, the enzyme enteropeptidase present in the intestinal mucosa activates trypsinogen by cleaving it to form trypsin; further cleavage results in chymotripsin.

Lower Gastrointestinal Tract

The lower gastrointestinal tract (GI), includes the small intestine and all of the large intestine. The intestine is also called the bowel or the gut. The lower GI starts at the pyloric sphincter of the stomach and finishes at the anus. The small intestine is subdivided into the duodenum, the jejunum and the ileum. The cecum marks the division between the small and large intestine. The large intestine includes the rectum and anal canal.

Small Intestine

Food starts to arrive in the small intestine one hour after it is eaten, and after two hours the stomach has emptied. Until this time the food is termed a bolus. It then becomes the partially digested semi-liquid termed chyme.

In the small intestine, the pH becomes crucial; it needs to be finely balanced in order to activate digestive enzymes. The chyme is very acidic, with a low pH, having been released from the stomach and needs to be made much more alkaline. This is achieved in the duodenum by the addition of bile from the gall bladder combined with the bicarbonate secretions from the pancreatic duct and also from secretions of bicarbonate-rich mucus from duodenal glands known as Brunner's glands. The chyme arrives in the intestines having been released from the stomach through the opening of the pyloric sphincter. The resulting alkaline fluid mix neutralises the gastric acid which would damage the lining of the intestine. The mucus component lubricates the walls of the intestine.

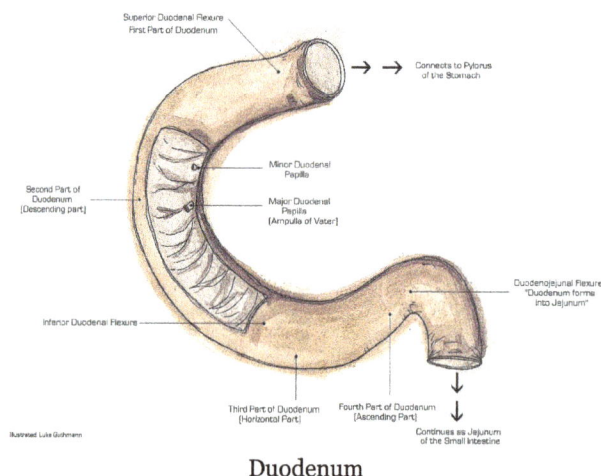

Duodenum

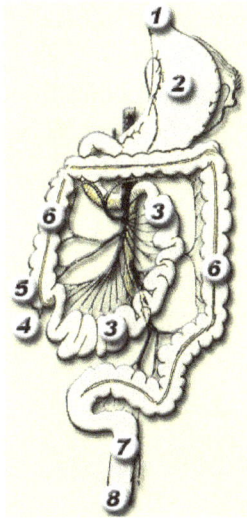

Lower GI tract - 3) Small intestine; 5) Cecum; 6) Large intestine

When the digested food particles are reduced enough in size and composition, they can be absorbed by the intestinal wall and carried to the bloodstream. The first receptacle for this chyme is the duodenal bulb. From here it passes into the first of the three sections of the small intestine, the duodenum. The next section is the jejunum and the third is the ileum. The duodenum is the first and shortest section of the small intestine. It is a hollow, jointed C-shaped tube connecting the stomach to the jejunum. It starts at the duodenal bulb and ends at the suspensory muscle of duodenum. The attachment of the suspensory muscle to the diaphragm is thought to help the passage of food by making a wider angle at its attachment.

Most food digestion takes place in the small intestine. Segmentation contractions act to mix and move the chyme more slowly in the small intestine allowing more time for absorption (and these continue in the large intestine). In the duodenum, pancreatic lipase is secreted together with a co-enzyme, colipase to further digest the fat content of the chyme. From this breakdown, smaller particles of emulsified fats called chylomicrons are produced. There are also digestive cells called enterocytes lining the intestines (the majority being in the small intestine). They are unusual cells in that they have villi on their surface which in turn have innumerable microvilli on their surface. All these villi make for a greater surface area, not only for the absorption of chyme but also for its further digestion by large numbers of digestive enzymes present on the microvilli.

The chylomicrons are small enough to pass through the enterocyte villi and into their lymph capillaries called lacteals. A milky fluid called chyle, consisting mainly of the emulsified fats of the chylomicrons, results from the absorbed mix with the lymph in the lacteals. Chyle is then transported through the lymphatic system to the rest of the body.

The suspensory muscle marks the end of the duodenum and the division between the upper gastrointestinal tract and the lower GI tract. The digestive tract continues as the jejunum which continues as the ileum. The jejunum, the midsection of the small intestine contains circular folds, flaps of doubled mucosal membrane which partially encircle and sometimes completely encircle the lumen of the intestine. These folds together with villi serve to increase the surface area of the jejunum enabling an increased absorption of digested sugars, amino acids and fatty acids into the bloodstream. The circular folds also slow the passage of food giving more time for nutrients to be absorbed.

The last part of the small intestine is the ileum. This also contains villi and vitamin B12; bile acids and any residue nutrients are absorbed here. When the chyme is exhausted of its nutrients the remaining waste material changes into the semi-solids called feces, which pass to the large intestine, where bacteria in the gut flora further break down residual proteins and starches.

Cecum

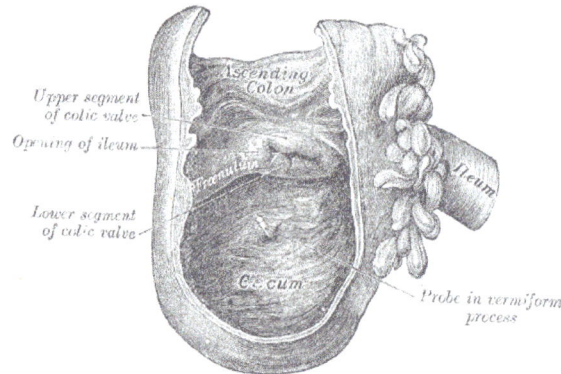

Cecum and beginning of ascending colon

The cecum is a pouch marking the division between the small intestine and the large intestine. The cecum receives chyme from the last part of the small intestine, the ileum, and connects to the ascending colon of the large intestine. At this junction there is a sphincter or valve, the ileocecal valve which slows the passage of chyme from the ileum, allowing further digestion. It is also the site of the appendix attachment.

Large Intestine

In the large intestine, the passage of the digesting food in the colon is a lot slower, taking from 12 to 50 hours until it is removed by defecation. The colon mainly serves as a site for the fermentation of digestible matter by the gut flora. The time taken varies considerably between individuals. The remaining semi-solid waste is termed feces and is removed by the coordinated contractions of the intestinal walls, termed peristalsis, which propels the excreta forward to reach the rectum and exit via defecation from the anus. The wall has an outer layer of longitudinal muscles, the taeniae coli, and an inner layer of circular muscles. The circular muscle keeps the material moving forward and also prevents any back flow of waste. Also of help in the action of peristalsis is the basal electrical rhythm that determines the frequency of contractions. The taeniae coli can be seen and are responsible for the bulges (haustra) present in the colon. Most parts of the GI tract are covered with serous membranes and have a mesentery. Other more muscular parts are lined with adventitia.

Blood Supply

The digestive system is supplied by the celiac artery. The celiac artery is the first major branch from the abdominal aorta, and is the only major artery that nourishes the digestive organs.

There are three main divisions – the left gastric artery, the common hepatic artery and the splenic artery.

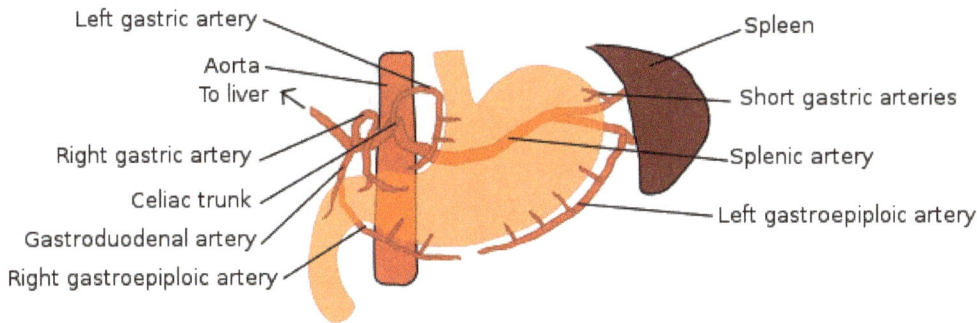

Blood supply to the digestive organs

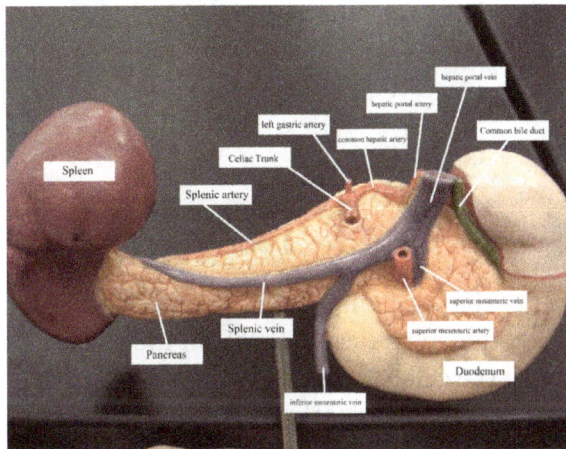

Arteries and veins around the pancreas and spleen

The celiac artery supplies the liver, stomach, spleen and the upper 1/3 of the duodenum (to the sphincter of Oddi) and the pancreas with oxygenated blood. Most of the blood is returned to the liver via the portal venous system for further processing and detoxification before returning to the systemic circulation via the hepatic veins.

The next branch from the abdominal aorta is the superior mesenteric artery, which supplies the regions of the digestive tract derived from the midgut, which includes the distal 2/3 of the duodenum, jejunum, ileum, cecum, appendix, ascending colon, and the proximal 2/3 of the transverse colon.

The final branch which is important for the digestive system is the inferior mesenteric artery, which supplies the regions of the digestive tract derived from the hindgut, which includes the distal 1/3 of the transverse colon, descending colon, sigmoid colon, rectum, and the anus above the pectinate line.

Nerve Supply

The enteric nervous system consists of some one hundred million neurons that are embedded in the peritoneum, the lining of the gastrointestinal tract extending from the esophagus to the anus. These neurons are collected into two plexuses - the myenteric (or Auerbach's) plexus that lies between the longitudinal and the smooth muscle layers, and the submucosal (or Meissner's) plexus that lies between the circular smooth muscle layer and the mucosa.

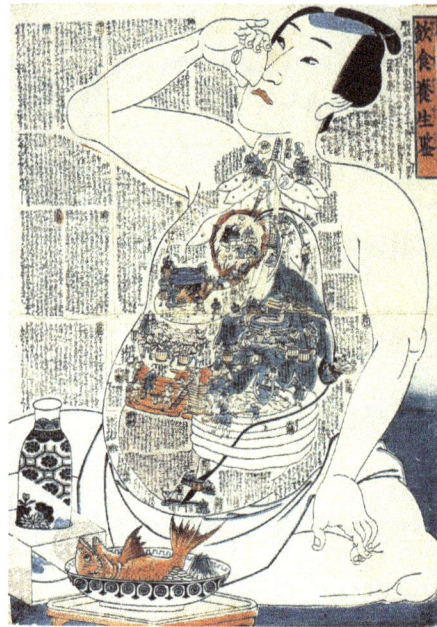

Dietary life rules, Japan, Edo period

Parasympathetic innervation to the ascending colon is supplied by the vagus nerve. Sympathetic innervation is supplied by the splanchnic nerves that join the celiac ganglia. Most of the digestive tract is innervated by the two large celiac ganglia, with the upper part of each ganglion joined by the greater splanchnic nerve and the lower parts joined by the lesser splanchnic nerve. It is from these ganglia that many of the gastric plexuses arise.

Development

Early in embryonic development, the embryo has three germ layers and abuts a yolk sac. During the second week of development, the embryo grows and begins to surround and envelop portions of this sac. The enveloped portions form the basis for the adult gastrointestinal tract. Sections of this foregut begin to differentiate into the organs of the gastrointestinal tract, such as the esophagus, stomach, and intestines.

During the fourth week of development, the stomach rotates. The stomach, originally lying in the midline of the embryo, rotates so that its body is on the left. This rotation also affects the part of the gastrointestinal tube immediately below the stomach, which will go on to become the duodenum. By the end of the fourth week, the developing duodenum begins to spout a small outpouching on its right side, the hepatic diverticulum, which will go on to become the biliary tree. Just below this is a second outpouching, known as the *cystic diverticulum*, that will eventually develop into the gallbladder.

Clinical Significance

Each part of the digestive system is subject to a wide range of disorders many of which can be congenital. Mouth diseases can also be caused by pathogenic bacteria, viruses, fungi and as a side effect of some medications. Mouth diseases include tongue diseases and salivary gland diseases. A common gum disease in the mouth is gingivitis which is caused by bacteria in plaque. The most common viral infection of the mouth is gingivostomatitis caused by herpes simplex. A common fungal infection is candidiasis commonly known as *thrush* which affects the mucous membranes of the mouth.

There are a number of esophageal diseases such as the development of Schatzki rings that can restrict the passageway, causing difficulties in swallowing. They can also completely block the esophagus.

Stomach diseases are often chronic conditions and include gastroparesis, gastritis, and peptic ulcers.

A number of problems including malnutrition and anemia can arise from malabsorption, the abnormal absorption of nutrients in the GI tract. Malabsorption can have many causes ranging from infection, to enzyme deficiencies such as exocrine pancreatic insufficiency. It can also arise as a result of other gastrointestinal diseases such as coeliac disease. Coeliac disease is an autoimmune disorder of the small intestine. This can cause vitamin deficiencies due to the improper absorption of nutrients in the small intestine. The small intestine can also be obstructed by a volvulus, a loop of intestine that becomes twisted enclosing its attached mesentery. This can cause mesenteric ischemia if severe enough.

A common disorder of the bowel is diverticulitis. Diverticula are small pouches that can form inside the bowel wall, which can become inflamed to give diverticulitis. This disease can have complications if an inflamed diverticulum bursts and infection sets in. Any infection can spread further to the lining of the abdomen (peritoneum) and cause potentially fatal peritonitis.

Crohn's disease is a common chronic inflammatory bowel disease (IBD), which can affect any part of the GI tract, but it mostly starts in the terminal ileum.

Ulcerative colitis an ulcerative form of colitis, is the other major inflammatory bowel disease which is restricted to the colon and rectum. Both of these IBDs can give an increased risk of the development of colorectal cancer. Ulcerative coliltis is the most common of the IBDs

Irritable bowel syndrome (IBS) is the most common of the functional gastrointestinal disorders. These are idiopathic disorders that the Rome process has helped to define.

Giardiasis is a disease of the small intestine caused by a protist parasite *Giardia lamblia*. This does not spread but remains confined to the lumen of the small intestine. It can often be asymptomatic, but as often can be indicated by a variety of symptoms. Giardiasis is the most common pathogenic parasitic infection in humans.

There are diagnostic tools mostly involving the ingestion of barium sulphate to investigate disorders of the GI tract. These are known as upper gastrointestinal series that enable imaging of the pharynx, larynx, oesophagous, stomach and small intestine and lower gastrointestinal series for imaging of the colon.

In Pregnancy

Gestation can predispose for certain digestive disorders. Gestational diabetes can develop in the mother as a result of pregnancy and while this often presents with few symptoms it can lead to pre-eclampsia.

Immune System

The immune system is the body's defense against infectious organisms and other invaders. Through a series of steps called the immune response, the immune system attacks organisms and substances that invade body systems and cause disease.

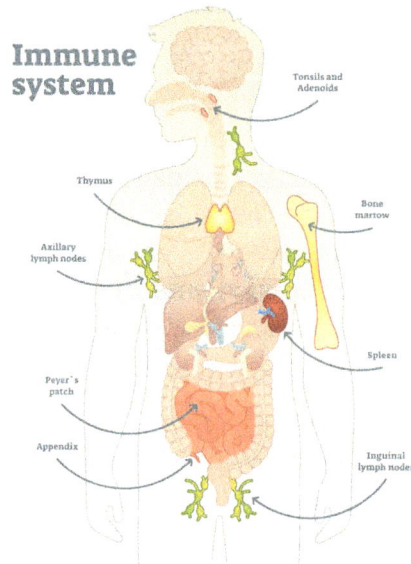

The immune system is made up of a network of cells, tissues, and organs that work together to protect the body. One of the important cells involved are white blood cells, also called leukocytes, which come in two basic types that combine to seek out and destroy disease-causing organisms or substances.

Leukocytes are produced or stored in many locations in the body, including the thymus, spleen, and bone marrow. For this reason, they're called the lymphoid organs. There are also clumps of lymphoid tissue throughout the body, primarily as lymph nodes, that house the leukocytes.

The leukocytes circulate through the body between the organs and nodes via lymphatic vessels and blood vessels. In this way, the immune system works in a coordinated manner to monitor the body for germs or substances that might cause problems.

The two basic types of leukocytes are:

1. Phagocytes, cells that chew up invading organisms

2. Lymphocytes, cells that allow the body to remember and recognize previous invaders and help the body destroy them

A number of different cells are considered phagocytes. The most common type is the neutrophil, which primarily fights bacteria. If doctors are worried about a bacterial infection, they might order a blood test to see if a patient has an increased number of neutrophils triggered by the infection. Other types of phagocytes have their own jobs to make sure that the body responds appropriately to a specific type of invader.

The two kinds of lymphocytes are B lymphocytes and T lymphocytes. Lymphocytes start out in the bone marrow and either stay there and mature into B cells, or they leave for the thymus gland, where they mature into T cells. B lymphocytes and T lymphocytes have separate functions: B lymphocytes are like the body's military intelligence system, seeking out their targets and sending defenses to lock onto them. T cells are like the soldiers, destroying the invaders that the intelligence system has identified.

Work of immune System

When antigens (foreign substances that invade the body) are detected, several types of cells work together to recognize them and respond. These cells trigger the B lymphocytes to produce antibodies, which are specialized proteins that lock onto specific antigens.

Once produced, these antibodies stay in a person's body, so that if his or her immune system encounters that antigen again, the antibodies are already there to do their job. So if someone gets sick with a certain disease, like chickenpox, that person usually won't get sick from it again.

This is also how immunizations prevent certain diseases. An immunization introduces the body to an antigen in a way that doesn't make someone sick, but does allow the body to produce antibodies that will then protect the person from future attack by the germ or substance that produces that particular disease.

Although antibodies can recognize an antigen and lock onto it, they are not capable of destroying it without help. That's the job of the T cells, which are part of the system that destroys antigens that have been tagged by antibodies or cells that have been infected or somehow changed. (Some T cells are actually called "killer cells.") T cells also are involved in helping signal other cells (like phagocytes) to do their jobs.

Antibodies also can neutralize toxins (poisonous or damaging substances) produced by different organisms. Lastly, antibodies can activate a group of proteins called complement that are also part of the immune system. Complement assists in killing bacteria, viruses, or infected cells.

All of these specialized cells and parts of the immune system offer the body protection against disease. This protection is called immunity.

Immunity

Humans have three types of immunity — innate, adaptive, and passive:

Innate Immunity

Everyone is born with innate (or natural) immunity, a type of general protection. Many of the germs that affect other species don't harm us. For example, the viruses that cause leukemia in cats or distemper in dogs don't affect humans. Innate immunity works both ways because some viruses that make humans ill — such as the virus that causes HIV/AIDS — don't make cats or dogs sick.

Innate immunity also includes the external barriers of the body, like the skin and mucous membranes (like those that line the nose, throat, and gastrointestinal tract), which are the first line of defense in preventing diseases from entering the body. If this outer defensive wall is broken (as through a cut), the skin attempts to heal the break quickly and special immune cells on the skin attack invading germs.

Adaptive Immunity

The second kind of protection is adaptive (or active) immunity, which develops throughout our lives. Adaptive immunity involves the lymphocytes and develops as people are exposed to diseases or immunized against diseases through vaccination.

Passive Immunity

Passive immunity is "borrowed" from another source and it lasts for a short time. For example, antibodies in a mother's breast milk give a baby temporary immunity to diseases the mother has been exposed to. This can help protect the baby against infection during the early years of childhood.

Everyone's immune system is different. Some people never seem to get infections, whereas others seem to be sick all the time. As people get older, they usually become immune to more germs as the immune system comes into contact with more and more of them. That's why adults and teens tend to get fewer colds than kids — their bodies have learned to recognize and immediately attack many of the viruses that cause colds.

Problems of the Immune System

Disorders of the immune system fall into four main categories:

1. Immunodeficiency disorders (primary or acquired)

2. Autoimmune disorders (in which the body's own immune system attacks its own tissue as foreign matter)

3. Allergic disorders (in which the immune system overreacts in response to an antigen)

4. Cancers of the immune system

Immunodeficiency Disorders

Immunodeficiencies happen when a part of the immune system is missing or not working properly. Some people are born with an immunodeficiency (known as primary immunodeficiencies), although symptoms of the disorder might not appear until later in life. Immunodeficiencies also can be acquired through infection or produced by drugs (these are sometimes called secondary immunodeficiencies).

Immunodeficiencies can affect B lymphocytes, T lymphocytes, or phagocytes. Examples of primary immunodeficiencies that can affect kids and teens are:

- IgA deficiency is the most common immunodeficiency disorder. IgA is an immunoglobulin that is found primarily in the saliva and other body fluids that help guard the entrances to the body. IgA deficiency is a disorder in which the body doesn't produce enough of the antibody IgA. People with IgA deficiency tend to have allergies or get more colds and other respiratory infections, but the condition is usually not severe.

- Severe combined immunodeficiency (SCID) is also known as the "bubble boy disease" after a Texas boy with SCID who lived in a germ-free plastic bubble. SCID is a serious immune system disorder that occurs because of a lack of both B and T lymphocytes, which makes it almost impossible to fight infections.

- DiGeorge syndrome (thymic dysplasia), a birth defect in which kids are born without a thymus gland, is an example of a primary T-lymphocyte disease. The thymus gland is where T lymphocytes normally mature.

- Chediak-Higashi syndrome and chronic granulomatous disease (CGD) both involve the inability of the neutrophils to function normally as phagocytes.

Acquired (or secondary) immunodeficiencies usually develop after someone has a disease, although they can also be the result of malnutrition, burns, or other medical problems. Certain medicines also can cause problems with the functioning of the immune system.

Acquired (secondary) immunodeficiencies include:

- HIV (human immunodeficiency virus) infection/AIDS (acquired immunodeficiency syndrome) is a disease that slowly and steadily destroys the immune system. It is caused by HIV, a virus that wipes out certain types of lymphocytes called T-helper cells. Without T-helper cells, the immune system is unable to defend the body against normally harmless organisms, which can cause life-threatening infections in people who have AIDS. Newborns can get HIV infection from their mothers while in the uterus, during the birth process, or during breastfeeding. People can get HIV infection by having unprotected sexual intercourse with an infected person or from sharing contaminated needles for drugs, steroids, or tattoos.

- Immunodeficiencies caused by medications. Some medicines suppress the immune system. One of the drawbacks of chemotherapy treatment for cancer, for example, is that it not only attacks cancer cells, but other fast-growing, healthy cells, including those found in the bone marrow and other parts of the immune system. In addition, people with autoimmune disorders or who have had organ transplants may need to take immunosuppressant medications, which also can reduce the immune system's ability to fight infections and can cause secondary immunodeficiency.

Autoimmune Disorders

In autoimmune disorders, the immune system mistakenly attacks the body's healthy organs and tissues as though they were foreign invaders. Autoimmune diseases include:

- Lupus: a chronic disease marked by muscle and joint pain and inflammation (the abnormal immune response also may involve attacks on the kidneys and other organs);

- Juvenile idiopathic arthritis: a disease in which the body's immune system acts as though certain body parts (such as the joints of the knee, hand, and foot) are foreign tissue and attacks them;

- Scleroderma: a chronic autoimmune disease that can lead to inflammation and damage of the skin, joints, and internal organs;

- Ankylosing spondylitis: a disease that involves inflammation of the spine and joints, causing stiffness and pain;

- Juvenile dermatomyositis: a disorder marked by inflammation and damage of the skin and muscles.

Allergic Disorders

Allergic disorders happen when the immune system overreacts to exposure to antigens in the environment. The substances that provoke such attacks are called allergens. The immune response can cause symptoms such as swelling, watery eyes, and sneezing, and even a life-threatening reaction called anaphylaxis. Medicines called antihistamines can relieve most symptoms.

Allergic disorders include:

- Asthma, a respiratory disorder that can cause breathing problems, often involves an allergic response by the lungs. If the lungs are oversensitive to certain allergens (like pollen, molds, animal dander, or dust mites), breathing tubes can become narrowed and swollen, making it hard for a person to breathe.

- Eczema is an itchy rash also known as atopic dermatitis. Although not necessarily caused by an allergic reaction, eczema most often happens in kids and teens who have allergies, hay fever, or asthma or who have a family history of these conditions.

- Allergies of several types can affect kids and teens. Environmental allergies (to dust mites, for example), seasonal allergies (such as hay fever), drug allergies (reactions to specific medications or drugs), food allergies (such as to nuts), and allergies to toxins (bee stings, for example) are the common conditions people usually refer to as allergies.

Cancers of the Immune System

Cancer happens when cells grow out of control. This can include cells of the immune system. Leukemia, which involves abnormal overgrowth of leukocytes, is the most common childhood cancer. Lymphoma involves the lymphoid tissues and is also one of the more common childhood cancers. With current treatments, most cases of both types of cancer in kids and teens are curable.

Lymphatic System

The lymphatic system is part of the immune system. It also maintains fluid balance and plays a role in absorbing fats and fat-soluble nutrients.

The lymphatic or lymph system involves an extensive network of vessels that passes through almost all our tissues to allow for the movement of a fluid called lymph. Lymph circulates through the body in a similar way to blood.

There are about 600 lymph nodes in the body. These nodes swell in response to infection, due to a build-up of lymph fluid, bacteria, or other organisms and immune system cells.

A person with a throat infection, for example, may feel that their "glands" are swollen. Swollen glands can be felt especially under the jaw, in the armpits, or in the groin area. These are, in fact, not glands but lymph nodes.

They should see a doctor if swelling does not go away, if nodes are hard or rubbery and difficult to move, if there is a fever, unexplained weight-loss, or difficulty breathing or swallowing.

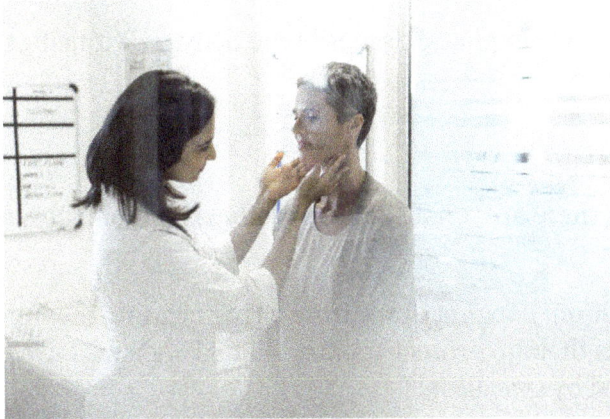

Lymph nodes, or "glands" may swell as the body responds to a threat

The lymphatic system has three main functions:

- It maintains the balance of fluid between the blood and tissues, known as fluid homeostasis.

- It forms part of the body's immune system and helps defend against bacteria and other intruders.

- It facilitates absorption of fats and fat-soluble nutrients in the digestive system.

The system has special small vessels called lacteals. These enable it to absorb fats and fat-soluble nutrients from the gut.

They work with the blood capillaries in the folded surface membrane of the small intestine. The blood capillaries absorb other nutrients directly into the bloodstream.

Anatomy

The lymphatic system consists of lymph vessels, ducts, nodes, and other tissues.

Around 2 liters of fluid leak from the cardiovascular system into body tissues every day. The lymphatic system is a network of vessels that collect these fluids, or lymph. Lymph is a clear fluid that is derived from blood plasma.

The lymph vessels form a network of branches that reach most of the body's tissues. They work in a similar way to the blood vessels. The lymph vessels work with the veins to return fluid from the tissues.

Unlike blood, the lymphatic fluid is not pumped but squeezed through the vessels when we use our muscles. The properties of the lymph vessel walls and the valves help control the movement of lymph. However, like veins, lymphatic vessels have valves inside them to stop fluid from flowing back in the wrong direction.

Lymph is drained progressively towards larger vessels until it reaches the two main channels, the lymphatic ducts in our trunk. From there, the filtered lymph fluid returns to the blood in the veins.

The vessels branch through junctions called lymph nodes. These are often referred to as glands, but they are not true glands as they do not form part of the endocrine system.

In the lymph nodes, immune cells assess for foreign material, such as bacteria, viruses, or fungus.

Lymph nodes are not the only lymphatic tissues in the body. The tonsils, spleen, and thymus gland are also lymphatic tissues.

The Function of Tonsils

In the back of the mouth, there are tonsils. These produce lymphocytes, a type of white blood cell, and antibodies.

They have a strategic position, hanging down from a ring forming the junction between the mouth and pharynx. This enables them to protect against inhaled and swallowed foreign bodies. The tonsils are the tissues affected by tonsillitis.

Spleen

The spleen is not connected to the lymphatic system in the same way as lymph nodes, but it is lymphoid tissue. This means it plays a role in the production of white blood cells that form part of the immune system.

Its other major role is to filter the blood to remove microbes and old and damaged red blood cells and platelets.

The Thymus Gland

The thymus gland is a lymphatic organ and an endocrine gland that is found just behind the sternum. It secretes hormones and is crucial in the production, maturation, and differentiation of immune T cells.

It is active in developing the immune system from before birth and through childhood.

The Bone Marrow

Bone marrow is not lymphatic tissue, but it can be considered part of the lymphatic system because it is here that the B cell lymphocytes of the immune system mature.

Liver of a Fetus

During gestation, the liver of a fetus is regarded as part of the lymphatic system as it plays a role in lymphocyte development.

Function

The lymph system has three main functions.

Fluid Balance

The lymphatic system helps maintain fluid balance. It returns excess fluid and proteins from the tissues that cannot be returned through the blood vessels.

The fluid is found in tissue spaces and cavities, in the tiny spaces surrounding cells, known as the interstitial spaces. These are reached by the smallest blood and lymph capillaries.

Around 90 percent of the plasma that reaches tissues from the arterial blood capillaries is returned by the venous capillaries and back along veins. The remaining 10 percent is drained back by the lymphatics.

Each day, around 2-3 liters is returned. This fluid includes proteins that are too large to be transported via the blood vessels.

Loss of the lymphatic system would be fatal within a day. Without the lymphatic system draining excess fluid, our tissues would swell, blood volume would be lost and pressure would increase.

Absorption

Most of the fats absorbed from the gastrointestinal tract are taken up in a part of the gut membrane in the small intestine that is specially adapted by the lymphatic system.

The lymphatic system has tiny lacteals in this part of the intestine that form part of the villi. These finger-like protruding structures are produced by the tiny folds in the absorptive surface of the gut.

Lacteals absorb fats and fat-soluble vitamins to form a milky white fluid called chyle.

This fluid contains lymph and emulsified fats, or free fatty acids. It delivers nutrients indirectly when it reaches the venous blood circulation. Blood capillaries take up other nutrients directly.

The Immune System

The third function is to defend the body against unwanted organisms. Without it, we would die very soon from an infection.

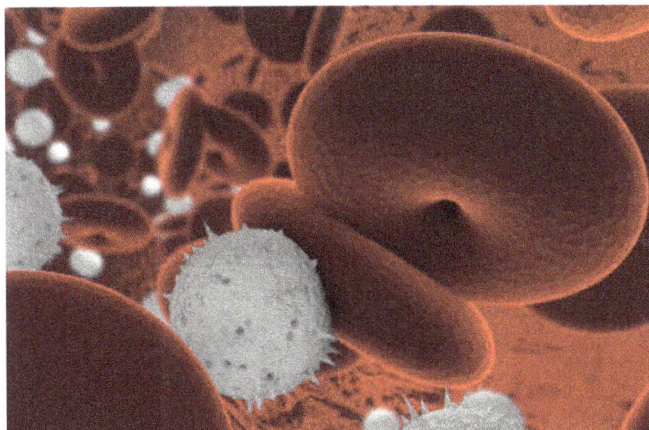

The lymphatic system produces white blood cells, or lymphocytes that are crucial in fending off infections

Our bodies are constantly exposed to potentially hazardous micro-organisms, such as infections.

The body's first line of defense involves:

- Physical barriers, such as the skin

- Toxic barriers, such as the acidic contents of the stomach

- "Friendly" bacteria in the body

However, pathogens often do succeed in entering the body despite these defenses. In this case, the lymphatic system enables our immune system to respond appropriately.

If the immune system is not able to fight off these micro-organisms, or pathogens, they can be harmful and even fatal.

A number of different immune cells and special molecules work together to fight off the unwanted pathogens.

Infection Fighting Mechanism of the Lymphatic System

The lymphatic system produces white blood cells, known as lymphocytes. There are two types of lymphocyte, T cells and B cells. They both travel through the lymphatic system.

As they reach the lymph nodes, they are filtered and become activated by contact with viruses, bacteria, foreign particles, and so on in the lymph fluid. From this stage, the pathogens, or invaders, are known as antigens.

As the lymphocytes become activated, they form antibodies and start to defend the body. They can also produce antibodies from memory if they have already encountered the specific pathogen in the past.

Collections of lymph nodes are concentrated in the neck, armpits, and groin. We become aware of these on one or both sides of the neck when we develop so-called "swollen glands" in response to an illness.

It is in the lymph nodes that the lymphocytes first encounter the pathogens, communicate with each other, and set off their defensive response.

Activated lymphocytes then pass further up the lymphatic system so that they can reach the bloodstream. Now, they are equipped to spread the immune response throughout the body, through the blood circulation.

The lymphatic system and the action of lymphocytes, of which the body has trillions, form part of what immunologists call the "adaptive immune response." These are highly specific and long-lasting responses to particular pathogens.

Diseases

The lymphatic system can stop working properly if nodes, ducts, vessels, or lymph tissues become blocked, infected, inflamed, or cancerous.

Lymphoma

Cancer that starts in the lymphatic system is known as lymphoma. It is the most serious lymphatic disease.

Hodgkin lymphoma affects a specific type of white blood cell known as Reed-Sternberg cells. Non-Hodgkin lymphoma refers to types that do not involve these cells.

Cancer that affects the lymphatic system is usually a secondary cancer. This means it has spread from a primary tumor, such as the breast, to nearby or regional lymph nodes.

Lymphadenitis

Sometimes, a lymph node swells because it becomes infected. The nodes may fill with pus, creating an abscess. The skin over the nodes may be red or streaky.

Localized lymphadenitis affects the nodes near the infection, for example, as a result of tonsilitis.

Generalized lymphadenitis can happen when a disease spreads through the bloodstream and affects the whole body. Causes range from sepsis to an upper respiratory tract infection.

Lymphedema

If the lymphatic system does not work properly, for example, if there is an obstruction, fluid may not drain effectively. As the fluid builds up, this can lead to swelling, for example in an arm or leg. This is lymphedema.

The skin may feel tight and hard, and skin problems may occur. In some cases, fluid may leak through the skin.

Obstruction can result from surgery, radiation therapy, injury, a condition known as lymphatic filariasis, or—rarely—a congenital disorder.

Swelling of Lymph Nodes

The "swollen glands," that occur, for example, in the neck during a throat infection, are in fact enlarged lymph nodes.

Lymph nodes can swell for two common reasons:

Reaction to an infection: The lymph nodes react when foreign material is presented to immune cells through the lymph that is drained from infected tissue.

Direct infection of the lymph nodes: The nodes can become infected and inflamed as a result of certain infections that need prompt antibiotic treatment. This is lymphadenitis.

Most people who have swollen glands with a cold or flu do not need to see a doctor.

However, medical advice should be sought if:

- Lymph nodes stay swollen for longer than 1 to 2 weeks

- A swollen lymph node feels hard or fixed in place

- Swelling is accompanied by fever, night sweats, or unexplained weight loss

Swollen lymph nodes can be symptoms of numerous conditions.

Glandular fever: Also known as infectious mononucleosis, or mono, this is a viral infection that can one cause longer-lasting swelling, a sore throat, and fatigue.

Tonsillitis: This is more common in children than in adults. It occurs when the lymph nodes at the back of the mouth are fighting infection, usually viral, but sometimes bacterial.

Pharyngitis: This bacterial infection is commonly called "strep throat." It is caused by group A streptococcus bacteria, and it can cause lymph nodes to swell.

Children are more prone to swollen lymph nodes because their immune systems are still developing their responses to infectious microbes.

Respiratory System

The respiratory system consists of all the organs involved in breathing. These include the nose, pharynx, larynx, trachea, bronchi and lungs. The respiratory system does two very important things: it brings oxygen into our bodies, which we need for our cells to live and function properly; and it helps us get rid of carbon dioxide, which is a waste product of cellular function. The nose, pharynx, larynx, trachea and bronchi all work like a system of pipes through which the air is funnelled down into our lungs. There, in very small air sacs called alveoli, oxygen is brought into the bloodstream and carbon dioxide is pushed from the blood out into the air. When something goes wrong with part of the respiratory system, such as an infection like pneumonia, it makes it harder for us to get the oxygen we need and to get rid of the waste product carbon dioxide. Common respiratory symptoms include breathlessness, cough, and chest pain.

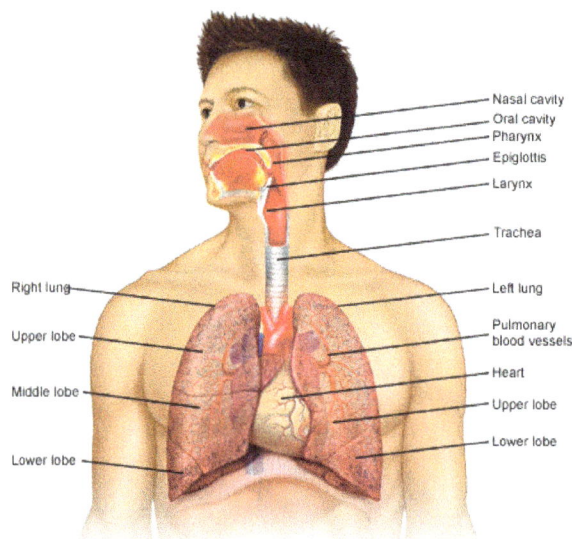

The Upper Airway and Trachea

When you breathe in, air enters your body through your nose or mouth. From there, it travels down your throat through the larynx (or voicebox) and into the trachea (or windpipe) before entering your lungs. All these structures act to funnel fresh air down from the outside world into your body. The upper airway is important because it must always stay open for you to be able to breathe. It also helps to moisten and warm the air before it reaches your lungs.

The Lungs

Structure

The lungs are paired, cone-shaped organs which take up most of the space in our chests, along with the heart. Their role is to take oxygen into the body, which we need for our cells to live and function properly, and to help us get rid of carbon dioxide, which is a waste product. We each have two lungs, a left lung and a right lung. These are divided up into 'lobes', or big sections of tissue separated by 'fissures' or dividers. The right lung has three lobes but the left lung has only two, because the heart takes up some of the space in the left side of our chest. The lungs can also be divided up into even smaller portions, called 'bronchopulmonary segments'.

These are pyramidal-shaped areas which are also separated from each other by membranes. There are about 10 of them in each lung. Each segment receives its own blood supply and air supply.

Working of Respiratory System

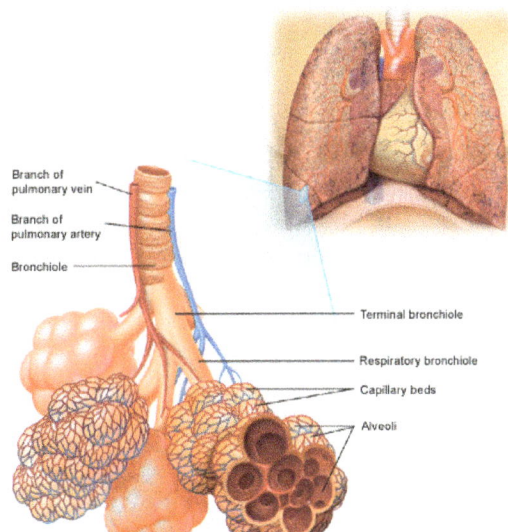

Air enters your lungs through a system of pipes called the bronchi. These pipes start from the bottom of the trachea as the left and right bronchi and branch many times throughout the lungs, until they eventually form little thin-walled air sacs or bubbles, known as the alveoli. The alveoli are where the important work of gas exchange takes place between the air and your blood. Covering each alveolus is a whole network of little blood vessel called capillaries, which are very small branches of the pulmonary arteries. It is important that the air in the alveoli and the blood in the capillaries are very close together, so that oxygen and carbon dioxide can move (or diffuse) between them. So, when you breathe in, air comes down the trachea and through the bronchi into the alveoli. This fresh air has lots of oxygen in it, and some of this oxygen will travel across the walls of the alveoli into your bloodstream. Travelling in the opposite direction is carbon dioxide, which crosses from the blood in the capillaries into the air in the alveoli and is then breathed out. In this way, you bring in to your body the oxygen that you need to live, and get rid of the waste product carbon dioxide.

Blood Supply

The lungs are very vascular organs, meaning they receive a very large blood supply. This is because the pulmonary arteries, which supply the lungs, come directly from the right side of your heart. They carry blood which is low in oxygen and high in carbon dioxide into your lungs so that the carbon dioxide can be blown off, and more oxygen can be absorbed into the bloodstream. The newly oxygen-rich blood then travels back through the paired pulmonary veins into the left side of your heart. From there, it is pumped all around your body to supply oxygen to cells and organs.

The Work of Breathing

The Pleurae

The lungs are covered by smooth membranes that we call pleurae. The pleurae have two layers, a 'visceral' layer which sticks closely to the outside surface of your lungs, and a 'parietal' layer which lines the inside of your chest wall (ribcage). The pleurae are important because they help you breathe in and out smoothly, without any friction. They also make sure that when your ribcage expands on breathing in, your lungs expand as well to fill the extra space.

The Diaphragm and Intercostal Muscles

When you breathe in (inspiration), your muscles need to work to fill your lungs with air. The diaphragm, a large, sheet-like muscle which stretches across your chest under the ribcage, does much of this work. At rest, it is shaped like a dome curving up into your chest. When you breathe in, the diaphragm contracts and flattens out, expanding the space in your chest and drawing air into your lungs. Other muscles, including the muscles between your ribs (the intercostal muscles) also help by moving your ribcage in and out. Breathing out (expiration) does not normally require your muscles to work. This is because your lungs are very elastic, and when your muscles relax at the end of inspiration your lungs simply recoil back into their resting position, pushing the air out as they go.

The Respiratory System through the Ages

Breathing for the Premature Baby

When a baby is born, it must convert from getting all of its oxygen through the placenta to absorbing oxygen through its lungs. This is a complicated process, involving many changes in both air and blood pressures in the baby's lungs. For a baby born preterm (before 37 weeks gestation), the change is even harder. This is because the baby's lungs may not yet be mature enough to cope with the transition. The major problem with a preterm baby's lungs is a lack of something called 'surfactant'. This is a substance produced by cells in the lungs which helps keep the air sacs, or alveoli, open. Without surfactant, the pressures in the lungs change and the smaller alveoli collapse.

This reduces the area across which oxygen and carbon dioxide can be exchanged, and not enough oxygen will be taken in. Normally, a foetus will begin producing surfactant from around 28-32 weeks gestation. When a baby is born before or around this age, it may not have enough surfactant to keep its lungs open. The baby may develop something called 'Neonatal Respiratory Distress Syndrome', or NRDS. Signs of NRDS include tachypnoea (very fast breathing), grunting, and cyanosis (blueness of the lips and tongue). Sometimes NRDS can be treated by giving the baby artificially made surfactant by a tube down into the baby's lungs.

The Respiratory System and Ageing

The normal process of ageing is associated with a number of changes in both the structure and function of the respiratory system. These include:

- Enlargement of the alveoli. The air spaces get bigger and lose their elasticity, meaning that there is less area for gases to be exchanged across. This change is sometimes referred to as 'senile emphysema'.

- The compliance (or springiness) of the chest wall decreases, so that it takes more effort to breathe in and out.

- The strength of the respiratory muscles (the diaphragm and intercostal muscles) decreases. This change is closely connected to the general health of the person.

All of these changes mean that an older person might have more difficulty coping with increased stress on their respiratory system, such as with an infection like pneumonia, than a younger person would.

Endocrine System

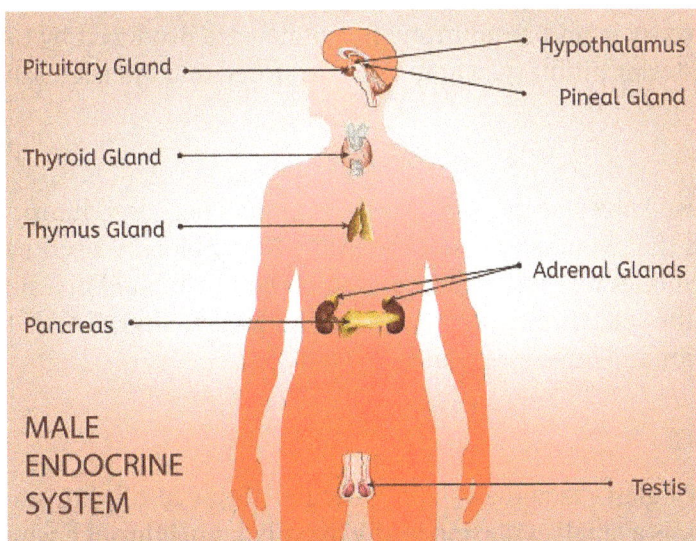

The endocrine system includes all of the glands of the body and the hormones produced by those glands. The glands are controlled directly by stimulation from the nervous system as well as by chemical receptors in the blood and hormones produced by other glands. By regulating the functions of organs in the body, these glands help to maintain the body's homeostasis. Cellular metabolism, reproduction, sexual development, sugar and mineral homeostasis, heart rate, and digestion are among the many processes regulated by the actions of hormones.

Anatomy of the Endocrine System

Hypothalamus

The hypothalamus is a part of the brain located superior and anterior to the brain stem and inferior to the thalamus. It serves many different functions in the nervous system, and is also responsible for the direct control of the endocrine system through the pituitary gland. The hypothalamus contains special cells called neurosecretory cells—neurons that secrete hormones:

- Thyrotropin-releasing hormone (TRH)

- Growth hormone-releasing hormone (GHRH)

- Growth hormone-inhibiting hormone (GHIH)

- Gonadotropin-releasing hormone (GnRH)

- Corticotropin-releasing hormone (CRH)

- Oxytocin

- Antidiuretic hormone (ADH)

All of the releasing and inhibiting hormones affect the function of the anterior pituitary gland. TRH stimulates the anterior pituitary gland to release thyroid-stimulating hormone. GHRH and GHIH work to regulate the release of growth hormone—GHRH stimulates growth hormone release, GHIH inhibits its release. GnRH stimulates the release of follicle stimulating hormone and luteinizing hormone while CRH stimulates the release of adrenocorticotropic hormone. The last two hormones—oxytocin and antidiuretic hormone—are produced by the hypothalamus and transported to the posterior pituitary, where they are stored and later released.

Pituitary Gland

The pituitary gland, also known as the hypophysis, is a small pea-sized lump of tissue connected to the inferior portion of the hypothalamus of the brain. Many blood vessels surround the pituitary gland to carry the hormones it releases throughout the body. Situated in a small depression in the sphenoid bone called the sella turcica, the pituitary gland is actually made of 2 completely separate structures: the posterior and anterior pituitary glands.

Posterior Pituitary

The posterior pituitary gland is actually not glandular tissue at all, but nervous tissue instead. The posterior pituitary is a small extension of the hypothalamus through which the axons of some of the neurosecretory cells of the hypothalamus extend. These neurosecretory cells create 2 hormones in the hypothalamus that are stored and released by the posterior pituitary:

- Oxytocin triggers uterine contractions during childbirth and the release of milk during breastfeeding.

- Antidiuretic hormone (ADH) prevents water loss in the body by increasing the re-uptake of water in the kidneys and reducing blood flow to sweat glands.

Anterior Pituitary

The anterior pituitary gland is the true glandular part of the pituitary gland. The function of the anterior pituitary gland is controlled by the releasing and inhibiting hormones of the hypothalamus. The anterior pituitary produces 6 important hormones:

- Thyroid stimulating hormone (TSH), as its name suggests, is a tropic hormone responsible for the stimulation of the thyroid gland.

- Adrenocorticotropic hormone (ACTH) stimulates the adrenal cortex, the outer part of the adrenal gland, to produce its hormones.

- Follicle stimulating hormone (FSH) stimulates the follicle cells of the gonads to produce gametes—ova in females and sperm in males.

- Luteinizing hormone (LH) stimulates the gonads to produce the sex hormones—estrogens in females and testosterone in males.

- Human growth hormone (HGH) affects many target cells throughout the body by stimulating their growth, repair, and reproduction.

- Prolactin (PRL) has many effects on the body, chief of which is that it stimulates the mammary glands of the breast to produce milk.

Pineal Gland

The pineal gland is a small pinecone-shaped mass of glandular tissue found just posterior to the thalamus of the brain. The pineal gland produces the hormone melatonin that helps to regulate the human sleep-wake cycle known as the circadian rhythm. The activity of the pineal gland is inhibited by stimulation from the photoreceptors of the retina. This light sensitivity causes melatonin to be produced only in low light or darkness. Increased melatonin production causes humans to feel drowsy at nighttime when the pineal gland is active.

Thyroid Gland

The thyroid gland is a butterfly-shaped gland located at the base of the neck and wrapped around the lateral sides of the trachea. The thyroid gland produces 3 major hormones:

- Calcitonin

- Triiodothyronine (T3)

- Thyroxine (T4)

Calcitonin is released when calcium ion levels in the blood rise above a certain set point. Calcitonin functions to reduce the concentration of calcium ions in the blood by aiding the absorption of calcium into the matrix of bones. The hormones T3 and T4 work together to regulate the body's metabolic rate. Increased levels of T3 and T4 lead to increased cellular activity and energy usage in the body.

Parathyroid Glands

The parathyroid glands are 4 small masses of glandular tissue found on the posterior side of the thyroid gland. The parathyroid glands produce the hormone parathyroid hormone (PTH), which is involved in calcium ion homeostasis. PTH is released from the parathyroid glands when calcium ion levels in the blood drop below a set point. PTH stimulates the osteoclasts to break down the calcium containing bone matrix to release free calcium ions into the bloodstream. PTH also triggers the kidneys to return calcium ions filtered out of the blood back to the bloodstream so that it is conserved.

Adrenal Glands

The adrenal glands are a pair of roughly triangular glands found immediately superior to the kidneys. The adrenal glands are each made of 2 distinct layers, each with their own unique functions: the outer adrenal cortex and inner adrenal medulla.

Adrenal Cortex

The adrenal cortex produces many cortical hormones in 3 classes: glucocorticoids, mineralocorticoids, and androgens.

- Glucocorticoids have many diverse functions, including the breakdown of proteins and lipids to produce glucose. Glucocorticoids also function to reduce inflammation and immune response.

- Mineralocorticoids, as their name suggests, are a group of hormones that help to regulate the concentration of mineral ions in the body.

- Androgens, such as testosterone, are produced at low levels in the adrenal cortex to regulate the growth and activity of cells that are receptive to male hormones. In adult males, the amount of androgens produced by the testes is many times greater than the amount produced by the adrenal cortex, leading to the appearance of male secondary sex characteristics.

Adrenal Medulla

The adrenal medulla produces the hormones epinephrine and norepinephrine under stimulation by the sympathetic division of the autonomic nervous system. Both of these hormones help to increase the flow of blood to the brain and muscles to improve the "fight-or-flight" response to stress. These hormones also work to increase heart rate, breathing rate, and blood pressure while decreasing the flow of blood to and function of organs that are not involved in responding to emergencies.

Pancreas

The pancreas is a large gland located in the abdominal cavity just inferior and posterior to the stomach. The pancreas is considered to be a heterocrine gland as it contains both endocrine and exocrine tissue. The endocrine cells of the pancreas make up just about 1% of the total mass of the pancreas and are found in small groups throughout the pancreas called islets of Langerhans. Within these islets are 2 types of cells—alpha and beta cells. The alpha cells produce the hormone glucagon, which is responsible for raising blood glucose levels. Glucagon triggers muscle and liver cells to break down the polysaccharide glycogen to release glucose into the bloodstream. The beta cells produce the hormone insulin, which is responsible for lowering blood glucose levels after a meal. Insulin triggers the absorption of glucose from the blood into cells, where it is added to glycogen molecules for storage.

Gonads

The gonads—ovaries in females and testes in males—are responsible for producing the sex hormones of the body. These sex hormones determine the secondary sex characteristics of adult females and adult males.

- *Testes*: The testes are a pair of ellipsoid organs found in the scrotum of males that produce the androgen testosterone in males after the start of puberty. Testosterone has effects on many parts of the body, including the muscles, bones, sex organs, and hair follicles. This hormone causes growth and increases in strength of the bones and muscles, including the accelerated growth of long bones during adolescence. During puberty, testosterone controls the growth and development of the sex organs and body hair of males, including pubic, chest, and facial hair. In men who have inherited genes for baldness testosterone triggers the onset of androgenic alopecia, commonly known as male pattern baldness.

- *Ovaries*: The ovaries are a pair of almond-shaped glands located in the pelvic body cavity lateral and superior to the uterus in females. The ovaries produce the female sex hormones progesterone and estrogens. Progesterone is most active in females during ovulation and pregnancy where it maintains appropriate conditions in the human body to support a developing fetus. Estrogens are a group of related hormones that function as the primary female sex hormones. The release of estrogen during puberty triggers the development of female secondary sex characteristics such as uterine development, breast development, and the growth of pubic hair. Estrogen also triggers the increased growth of bones during adolescence that lead to adult height and proportions.

Thymus

The thymus is a soft, triangular-shaped organ found in the chest posterior to the sternum. The thymus produces hormones called thymosins that help to train and develop T-lymphocytes during fetal development and childhood. The T-lymphocytes produced in the thymus go on to protect the body from pathogens throughout a person's entire life. The thymus becomes inactive during puberty and is slowly replaced by adipose tissue throughout a person's life.

Other Hormone Producing Organs

In addition to the glands of the endocrine system, many other non-glandular organs and tissues in the body produce hormones as well.

- *Heart*: The cardiac muscle tissue of the heart is capable of producing the hormone atrial natriuretic peptide (ANP) in response to high blood pressure levels. ANP works to reduce blood pressure by triggering vasodilation to provide more space for the blood to travel through. ANP also reduces blood volume and pressure by causing water and salt to be excreted out of the blood by the kidneys.

- *Kidneys*: The kidneys produce the hormone erythropoietin (EPO) in response to low levels of oxygen in the blood. EPO released by the kidneys travels to the red bone marrow where it stimulates an increased production of red blood cells. The number of red blood cells increases the oxygen carrying capacity of the blood, eventually ending the production of EPO.

- *Digestive System*: The hormones cholecystokinin (CCK), secretin, and gastrin are all produced by the organs of the gastrointestinal tract. CCK, secretin, and gastrin all help to regulate the secretion of pancreatic juice, bile, and gastric juice in response to the presence of food in the stomach. CCK is also instrumental in the sensation of satiety or "fullness" after eating a meal.

- *Adipose*: Adipose tissue produces the hormone leptin that is involved in the management of appetite and energy usage by the body. Leptin is produced at levels relative to the amount of adipose tissue in the body, allowing the brain to monitor the body's energy storage condition. When the body contains a sufficient level of adipose for energy storage, the level of leptin in the blood tells the brain that the body is not starving and may work normally. If the level of adipose or leptin decreases below a certain threshold, the body enters starvation mode and attempts to conserve energy through increased hunger and food intake and decreased energy usage. Adipose tissue also produces very low levels of estrogens in both men and women. In obese people the large volume of adipose tissue may lead to abnormal estrogen levels.

- *Placenta*: In pregnant women, the placenta produces several hormones that help to maintain pregnancy. Progesterone is produced to relax the uterus, protect the fetus from the mother's immune system, and prevent premature delivery of the fetus. Human chorionic gonadotropin (HCG) assists progesterone by signaling the ovaries to maintain the production of estrogen and progesterone throughout pregnancy.

- *Local Hormones*: Prostaglandins and leukotrienes are produced by every tissue in the body (except for blood tissue) in response to damaging stimuli. These two hormones mainly affect the cells that are local to the source of damage, leaving the rest of the body free to function normally.

  1. Prostaglandins cause swelling, inflammation, increased pain sensitivity, and increased local body temperature to help block damaged regions of the body from infection or further damage. They act as the body's natural bandages to keep pathogens out and swell around damaged joints like a natural cast to limit movement.

  2. Leukotrienes help the body heal after prostaglandins have taken effect by reducing inflammation while helping white blood cells to move into the region to clean up pathogens and damaged tissues.

Physiology of the Endocrine System

Endocrine System vs. Nervous System Function

The endocrine system works alongside of the nervous system to form the control systems of the body. The nervous system provides a very fast and narrowly targeted system to turn on specific glands and muscles throughout the body. The endocrine system, on the other hand, is much slower acting, but has very widespread, long lasting, and powerful effects. Hormones are distributed by glands through the bloodstream to the entire body, affecting any cell with a receptor for a particular hormone. Most hormones affect cells in several organs or throughout the entire body, leading to many diverse and powerful responses.

Hormone Properties

Once hormones have been produced by glands, they are distributed through the body via the bloodstream. As hormones travel through the body, they pass through cells or along the plasma membranes of cells until they encounter a receptor for that particular hormone. Hormones can only affect target cells that have the appropriate receptors. This property of hormones is known as specificity. Hormone specificity explains how each hormone can have specific effects in widespread parts of the body.

Many hormones produced by the endocrine system are classified as tropic hormones. A tropic hormone is a hormone that is able to trigger the release of another hormone in another gland. Tropic hormones provide a pathway of control for hormone production as well as a way for glands to be controlled in distant regions of the body. Many of the hormones produced by the pituitary gland, such as TSH, ACTH, and FSH are tropic hormones.

Hormonal Regulation

The levels of hormones in the body can be regulated by several factors. The nervous system can control hormone levels through the action of the hypothalamus and its releasing and inhibiting hormones. For example, TRH produced by the hypothalamus stimulates the anterior pituitary to produce TSH. Tropic hormones provide another level of control for the release of hormones. For example, TSH is a tropic hormone that stimulates the thyroid gland to produce T3 and T4. Nutrition can also control the levels of hormones in the body. For example, the thyroid hormones T3 and T4 require 3 or 4 iodine atoms, respectively, to be produced. In people lacking iodine in their diet, they will fail to produce sufficient levels of thyroid hormones to maintain a healthy metabolic rate. Finally, the number of receptors present in cells can be varied by cells in response to hormones. Cells that are exposed to high levels of hormones for extended periods of time can begin to reduce the number of receptors that they produce, leading to reduced hormonal control of the cell.

Classes of Hormones

Hormones are classified into 2 categories depending on their chemical make-up and solubility: water-soluble and lipid-soluble hormones. Each of these classes of hormones has specific mechanisms for their function that dictate how they affect their target cells.

- *Water-soluble hormones*: Water-soluble hormones include the peptide and amino-acid hormones such as insulin, epinephrine, HGH, and oxytocin. As their name indicates, these hormones are soluble in water. Water-soluble hormones are unable to pass through the phospholipid bilayer of the plasma membrane and are therefore dependent upon receptor molecules on the surface of cells. When a water-soluble hormone binds to a receptor molecule on the surface of a cell, it triggers a reaction inside of the cell. This reaction may change a factor inside of the cell such as the permeability of the membrane or the activation of another molecule. A common reaction is to cause molecules of cyclic adenosine monophosphate (cAMP) to be synthesized from adenosine triphosphate (ATP) present in the cell. cAMP acts as a second messenger within the cell where it binds to a second receptor to change the function of the cell's physiology.

- *Lipid-soluble hormones*: Lipid-soluble hormones include the steroid hormones such as testosterone, estrogens, glucocorticoids, and mineralocorticoids. Because they are soluble in lipids, these hormones are able to pass directly through the phospholipid bilayer of the plasma membrane and bind directly to receptors inside the cell nucleus. Lipid-soluble hormones are able to directly control the function of a cell from these receptors, often triggering the transcription of particular genes in the DNA to produce "messenger RNAs (mRNAs)" that are used to make proteins that affect the cell's growth and function.

Integumentary System

The integumentary system is the set of organs that forms the external covering of the body and protects it from many threats such as infection, desiccation, abrasion, chemical assault and radiation damage. IN humans the integumentary system includes the skin – a thickened keratinized epithelium made of multiple layers of cells that is largely impervious to water. It also contains specialized cells that secrete melanin to protect the body from the carcinogenic effects of UV rays and cells that have an immune function. Sweat glands that excrete wastes and regulate body temperature are also part of the integumentary system. Somatosensory receptors and nociceptors are important components of this organ system that serve as warning sensors, allowing the body to move away from noxious stimuli.

Organs of the Integumentary System

The skin consists of two layers – the dermis and the epidermis. Together, these two layers form the largest organ in the body, with a surface area of nearly 2 square meters.

The epidermis is the outer layer, resting atop the dermis. There is no direct blood supply to the epidermis and therefore, the cells of this stratified squamous tissue obtain nutrients and oxygen through diffusion. This layer also cushions underlying tissues and protects them from desiccation. In hot, dry environments, water is first lost from this layer. Similarly, extended exposure to water during baths or during swimming, crinkles the skin since water is absorbed and retained in the epidermis.

The epidermis is made of four layers – the stratum basale, stratum spinosum, stratum granulosum and stratum corneum. In each of these layers, keratinocytes undergo successive steps in differentiation beginning with the proliferative layer in the innermost stratum basale containing keratinocyte stem cells. After division, cells migrate outwards to form a layer of spiny cells called stratum spinosum. The nuclei of these cells are primarily involved in transcribing large amounts of keratin mRNA and other microfibrils that form impermeable cell junctions. The next layer of the epidermis is called stratum granulosum and contains keratinocytes with a granular cytoplasm. This stage in keratinocyte maturation is characterized by the formation of the lipid barrier of the body. The presence of keratohyalin granules is important for crosslinking keratin filaments and dehydrating cells to form tight, interlinked layers of cells that perform the barrier function of skin. The outermost layer is called the stratum corneum and is directly exposed to the external environment. It consists of multiple layers of terminally differentiated keratinocytes that are also called corneocytes. These cells do not have a nucleus and contain copious amounts of keratin filaments. This layer of the epidermis provides mechanical strength and rigidity to the structure of skin. These anucleated cells are resistant to virus attack and are replaced every 15 days, preventing them from becoming a reservoir of infection. The parts of the skin that have no hair follicles have an extra layer of epithelium called the stratum lucidum that is sandwiched between the stratum granulosum and stratum corneum. This extra layer makes the epithelium of these regions 'thicker' than those in other parts of the body. Usually, this is the skin on the palms of the hands and soles of the feet, and in addition to stratum lucidum, is also well supplied with nerve endings.

The second major section of the integument is the dermis, and is occasionally called the 'true skin' since it is supplied with blood vessels and nerve endings. Sebaceous glands and sweat glands are also present in the dermis. The closest that the dermis gets to the external environment is at structures called dermal papillae. These are finger-like projections into the epidermis and, on the palms, form fingerprints.

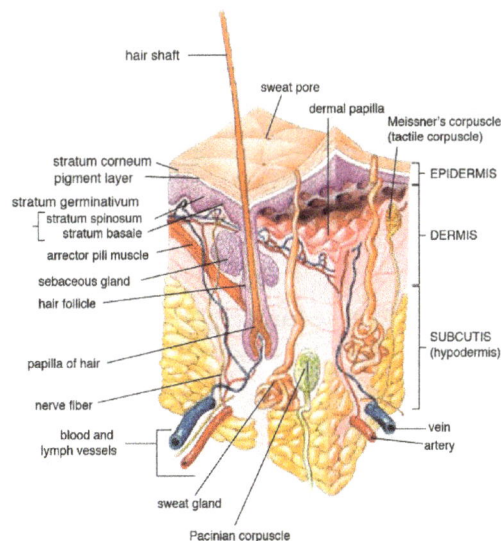

Image shows cross section of skin, with various dermal and epidermal layers, glands, nerves and blood vessels

Sebaceous glands produce sebum – an oily, waxy secretion containing many lipids. The cells forming a sebaceous gland have extremely short lifespans – barely over a week. The soles of the feet are free from sebaceous glands, though the sections of skin between the toes is richly supplied with these structures. Sebum also forms a part of ear wax. These lipids can provide a rich environment for the growth of bacteria, and therefore contribute towards body odor, either when the glands are clogged or when the sebum is not removed periodically.

The dermis also plays host to sweat glands. Sweat, in contrast to sebum, is a water-based secretion, containing electrolytes – sodium salts, urea, and even trace amounts of uric acid. While most water soluble waste products are removed in the urine, sweat also contributes towards clearing some of the metabolic byproducts of the body. The presence of many acids, such as lactic acid and acetic acid, makes sweat mildly acidic. A subsection of sweat glands, called apocrine glands, even release proteins, carbohydrates, lipids or steroids. Sweat from these glands, along with sebum, can encourage bacterial growth, and form the site for infection, odor or rashes.

Functions of the Integumentary System

Each layer of the skin contributes to the overall function within the body. The most obvious role of the skin is to protect the body from external aggression.

Barrier Function

While the skin may seem like a delicate organ, its stupendous role becomes apparent after an injury removes the skin from a region. In fact, preventing infections and regulating body temperature are major challenges in burn victims. Layers of tightly bound, heavily keratinized, anucleated cells provide the first line of defense by forming a physical barrier. The mildly acidic nature of skin secretions also contributes towards preventing pathogenic colonization. Lipids secreted by the skin are another chemical barrier, preventing the loss of water, especially in dry or hot environments. Alternatively, the skin also prevents the body from bloating in an hypotonic environment. Finally, the integumentary system contains resident immune cells that are adept at clearing minor infections.

Thermoregulation

Sweat glands are necessary for thermoregulation, whether it is while working up a sweat during exercise or breaking a fever. Sweat allows the body to cool down. On the other hand, goosebumps arising from the contraction of arrector pili muscles can keep the body warm, especially in hairy mammals.

Excretion

Sweat and sebum also have an excretory role for water and fat soluble metabolites respectively. For instance, excess vitamin B from supplements is removed through urine and sweat.

Sensation and Chemical Synthesis

Nerve endings on the skin help in sensing touch, pressure, heat, cold as well as the nature and intensity of damaging stimuli. The skin is also necessary for the production of melanin that prevents damage from UV rays – whether it is a sunburn or skin cancer. Upon exposure to the sun, in addition to melanin production, the skin also synthesizes vitamin D that contributes to bone health and enhances bone density.

Urinary System

The urinary system consists of the kidneys, ureters, urinary bladder, and urethra. The kidneys filter the blood to remove wastes and produce urine. The ureters, urinary bladder, and urethra together form the urinary tract, which acts as a plumbing system to drain urine from the kidneys, store it, and then release it during urination. Besides filtering and eliminating wastes from the body, the urinary system also maintains the homeostasis of water, ions, pH, blood pressure, calcium and red blood cells.

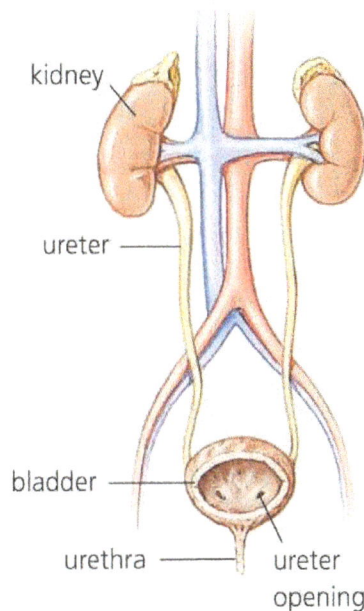

Urinary System Anatomy

Kidneys

The kidneys are a pair of bean-shaped organs found along the posterior wall of the abdominal cavity. The left kidney is located slightly higher than the right kidney because the right side of the liver is much larger than the left side. The kidneys, unlike the other organs of the abdominal cavity, are located posterior to the peritoneum and touch the muscles of the back. The kidneys are surrounded by a layer of adipose that holds them in place and protects them from physical damage. The kidneys filter metabolic wastes, excess ions, and chemicals from the blood to form urine.

Ureters

The ureters are a pair of tubes that carry urine from the kidneys to the urinary bladder. The ureters are about 10 to 12 inches long and run on the left and right sides of the body parallel to the vertebral column. Gravity and peristalsis of smooth muscle tissue in the walls of the ureters move urine toward the urinary bladder. The ends of the ureters extend slightly into the urinary bladder and are sealed at the point of entry to the bladder by the ureterovesical valves. These valves prevent urine from flowing back towards the kidneys.

Urinary Bladder

The urinary bladder is a sac-like hollow organ used for the storage of urine. The urinary bladder is located along the body's midline at the inferior end of the pelvis. Urine entering the urinary bladder from the ureters slowly fills the hollow space of the bladder and stretches its elastic walls. The walls of the bladder allow it to stretch to hold anywhere from 600 to 800 milliliters of urine.

Urethra

The urethra is the tube through which urine passes from the bladder to the exterior of the body. The female urethra is around 2 inches long and ends inferior to the clitoris and superior to the vaginal opening. In males, the urethra is around 8 to 10 inches long and ends at the tip of the penis. The urethra is also an organ of the male reproductive system as it carries sperm out of the body through the penis.

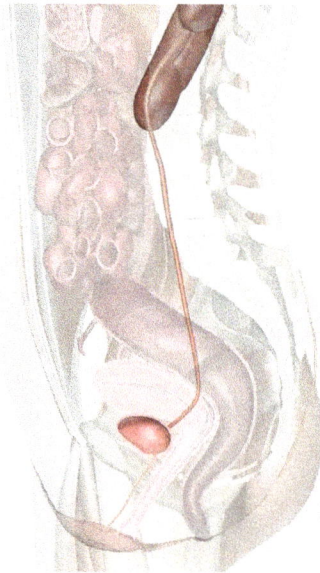

The flow of urine through the urethra is controlled by the internal and external urethral sphincter muscles. The internal urethral sphincter is made of smooth muscle and opens involuntarily when the bladder reaches a certain set level of distention. The opening of the internal sphincter results in the sensation of needing to urinate. The external urethral sphincter is made of skeletal muscle and may be opened to allow urine to pass through the urethra or may be held closed to delay urination.

Urinary System Physiology

Maintenance of Homeostasis

The kidneys maintain the homeostasis of several important internal conditions by controlling the excretion of substances out of the body.

Ions

The kidney can control the excretion of potassium, sodium, calcium, magnesium, phosphate, and chloride ions into urine. In cases where these ions reach a higher than normal concentration, the kidneys can increase their excretion out of the body to return them to a normal level. Conversely, the kidneys can conserve these ions when they are present in lower than normal levels by allowing the ions to be reabsorbed into the blood during filtration.

pH

The kidneys monitor and regulate the levels of hydrogen ions (H+) and bicarbonate ions in the blood to control blood pH. H+ ions are produced as a natural byproduct of the metabolism of dietary proteins and accumulate in the blood over time. The kidneys excrete excess H+ ions into urine for elimination from the body. The kidneys also conserve bicarbonate ions, which act as important pH buffers in the blood.

Osmolarity

The cells of the body need to grow in an isotonic environment in order to maintain their fluid and electrolyte balance. The kidneys maintain the body's osmotic balance by controlling the amount of water that is filtered out of the blood and excreted into urine. When a person consumes a large amount of water, the kidneys reduce their reabsorption of water to allow the excess water to be excreted in urine. This results in the production of dilute, watery urine. In the case of the body being dehydrated, the kidneys reabsorb as much water as possible back into the blood to produce highly concentrated urine full of excreted ions and wastes. The changes in excretion of water are controlled by antidiuretic hormone (ADH). ADH is produced in the hypothalamus and released by the posterior pituitary gland to help the body retain water.

Blood Pressure

The kidneys monitor the body's blood pressure to help maintain homeostasis. When blood pressure is elevated, the kidneys can help to reduce blood pressure by reducing the volume of blood in the body. The kidneys are able to reduce blood volume by reducing the reabsorption of water into the blood and producing watery, dilute urine. When blood pressure becomes too low, the kidneys can produce the enzyme renin to constrict blood vessels and produce concentrated urine, which allows more water to remain in the blood.

Filtration

Inside each kidney are around a million tiny structures called nephrons. The nephron is the functional unit of the kidney that filters blood to produce urine. Arterioles in the kidneys deliver blood to a bundle of capillaries surrounded by a capsule called a glomerulus. As blood flows through the glomerulus, much of the blood's plasma is pushed out of the capillaries and into the capsule, leaving the blood cells and a small amount of plasma to continue flowing through the capillaries. The liquid filtrate in the capsule flows through a series of tubules lined with filtering cells and surrounded by capillaries. The cells surrounding the tubules selectively absorb water and substances from the filtrate in the tubule and return it to the blood in the capillaries. At the same time, waste products present in the blood are secreted into the filtrate. By the end of this process, the filtrate in the tubule has become urine containing only water, waste products, and excess ions. The blood exiting the capillaries has reabsorbed all of the nutrients along with most of the water and ions that the body needs to function.

Storage and Excretion of Wastes

After urine has been produced by the kidneys, it is transported through the ureters to the urinary bladder. The urinary bladder fills with urine and stores it until the body is ready for its excretion. When the volume of the urinary bladder reaches anywhere from 150 to 400 milliliters, its walls begin to stretch and stretch receptors in its walls send signals to the brain and spinal cord. These signals result in the relaxation of the involuntary internal urethral sphincter and the sensation of needing to urinate. Urination may be delayed as long as the bladder does not exceed its maximum volume, but increasing nerve signals lead to greater discomfort and desire to urinate.

Urination is the process of releasing urine from the urinary bladder through the urethra and out of the body. The process of urination begins when the muscles of the urethral sphincters relax, allowing urine to pass through the urethra. At the same time that the sphincters relax, the smooth muscle in the walls of the urinary bladder contract to expel urine from the bladder.

Production of Hormones

The kidneys produce and interact with several hormones that are involved in the control of systems outside of the urinary system.

Calcitriol

Calcitriol is the active form of vitamin D in the human body. It is produced by the kidneys from precursor molecules produced by UV radiation striking the skin. Calcitriol works together with parathyroid hormone (PTH) to raise the level of calcium ions in the bloodstream. When the level of calcium ions in the blood drops below a threshold level, the parathyroid glands release PTH, which in turn stimulates the kidneys to release calcitriol. Calcitriol promotes the small intestine to absorb calcium from food and deposit it into the bloodstream. It also stimulates the osteoclasts of the skeletal system to break down bone matrix to release calcium ions into the blood.

Erythropoietin

Erythropoietin, also known as EPO, is a hormone that is produced by the kidneys to stimulate the production of red blood cells. The kidneys monitor the condition of the blood that passes through their capillaries, including the oxygen-carrying capacity of the blood. When the blood becomes hypoxic, meaning that it is carrying deficient levels of oxygen, cells lining the capillaries begin producing EPO and release it into the bloodstream. EPO travels through the blood to the red bone marrow, where it stimulates hematopoietic cells to increase their rate of red blood cell production. Red blood cells contain hemoglobin, which greatly increases the blood's oxygen-carrying capacity and effectively ends the hypoxic conditions.

Renin

Renin is not a hormone itself, but an enzyme that the kidneys produce to start the renin-angiotensin system (RAS). The RAS increases blood volume and blood pressure in response to low blood pressure, blood loss, or dehydration. Renin is released into the blood where it catalyzes angiotensinogen from the liver into angiotensin I. Angiotensin I is further catalyzed by another enzyme into Angiotensin II.

Angiotensin II stimulates several processes, including stimulating the adrenal cortex to produce the hormone aldosterone. Aldosterone then changes the function of the kidneys to increase the reabsorption of water and sodium ions into the blood, increasing blood volume and raising blood pressure. Negative feedback from increased blood pressure finally turns off the RAS to maintain healthy blood pressure levels.

Excretory System

The human excretory system functions to remove waste from the human body. This system consists of specialized structures and capillary networks that assist in the excretory process. The human excretory system includes the kidneys and their functional unit, the nephron. The excretory activity of the kidneys is modulated by specialized hormones that regulate the amount of absorption within the nephron.

Parts of the Human Excretory System

Every living organism generates waste in its body and has a mechanism to expel it. In humans, waste generation and disposal are taken care of by the human excretory system. The human excretory system comprises of the following structures:

- 2 Kidneys
- 2 Ureters
- 1 Urinary bladder
- 1 Urethra

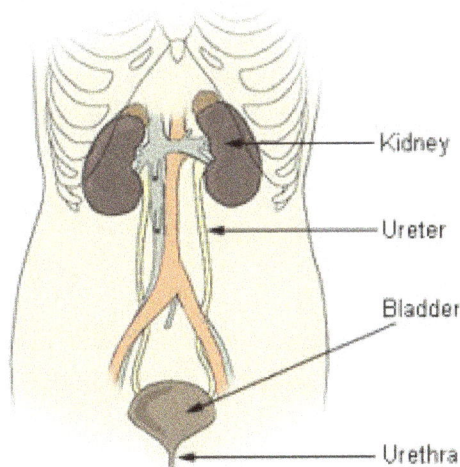

Components of the urinary system

Kidneys

Kidneys are the main organ of the human excretory system. The kidneys are paired organs in each individual. They are the primary excretory organ in humans and are located one on each side of the spine at the level of the liver. They are divided into three regions- the renal cortex which is the outer layer, the renal medulla which is the inner layer and the renal pelvis which is responsible for carrying the urine from the kidney to the ureter. The functional unit of a kidney is called the nephron.

Ureters

There is one ureter that comes out of each kidney as an extension of the renal pelvis. The ureter is a thin muscular tube that carries urine from the kidneys to the bladder.

Urinary Bladder

It is a sac-like structure that is lined with smooth muscle layer and is responsible for storage of urine till it is expelled from the body by micturition. Micturition is the act of expelling urine from the body. The bladder receives urine from the ureters, one from each kidney. The level of the urinary bladder placement in the body differs in men and women.

Urethra

This is a tube that arises from the urinary bladder and functions to expel urine to the outside by micturition. The urethra is shorter in females and longer in the males. In males, the urethra functions as a common path for sperms and urine. The opening of the urethra is guarded by a sphincter that is autonomically controlled.

Other Excretory Organs

Apart from the above mentioned excretory organs, there are other organs that also perform some form of excretion.

Skin

The skin is the largest organ in the body. Its primary function is to protect the different organs of the body. However, the skin helps in excretion by the way of sweat. The skin eliminates compounds like NaCl, some amount of urea etc.

Lungs

Lungs are the primary respiratory organs and they help take in oxygen and expel carbon dioxide. But, in this process, they also function to eliminate some amount of water in the form of vapour.

Liver

The liver has an important function in excretion. It is said to be the first line of defence when it comes to hormones, fats, alcohol, and drugs. Most drugs undergo a first pass metabolism which occurs in the liver. Few drugs are eliminated directly by the kidneys. The liver is said to play a role in the elimination of excess fats and cholesterol that is essential to the health of the body.

Structure of a Nephron

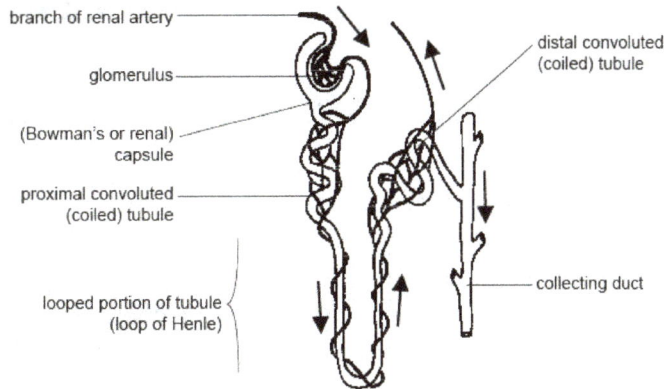

The structural and functional unit of the kidney is the nephrons. Each kidney consists of millions of nephrons that are all functioning together to filter urine and expel the waste products. Each kidney consists of the following parts:

- Bowman's capsule– is the first part of the nephron which is a cup-shaped structure and receives the blood vessels. The glomerular filtration occurs here. The blood cells and proteins remain in the blood.

- Proximal Convoluted Tubule– The Bowman's capsule extends downwards to form the proximal tubule. Water and reusable materials from the blood are now reabsorbed back into it.

- The loop of Henle– The proximal tubule leads to the formation of a u-shaped loop called the Loop of Henle. The Loop of Henle has three parts: The descending limb, the u-shaped bend, and the ascending limb. It is in this area that the urine becomes concentrated as water is reabsorbed. The descending limb is freely permeable to water whereas the ascending limb is impermeable to it.

- Distal Convoluted Tubule– The Loop of Henle leads into the distal convoluted tubule which is where the kidney hormones cause their effect. And the distal convoluted tubule leads to the collecting ducts.

- Collecting Duct– The distal convoluted tubule of each nephron leads to the collecting ducts. The collecting ducts together form the renal pelvis through which the urine passes into the ureter and then into the urinary bladder.

Functions of the Excretory System

The excretory system performs many functions such as:

- Helps eliminate waste products such as urea, uric acid ammonia, and other products via urine.

- It helps maintain the osmotic level of blood and plasma.

- It helps maintain the electrolyte balance in the body.

- And it also helps in the metabolism of those drugs that do not get metabolized in the liver.

Chapter 3

Human Evolution, Genetics and Reproduction

Human genetics delves into the study of the inheritable traits which have been passed on through generations. Evolution and reproduction are two significant aspects of this field which aid the understanding of human biology. The fundamental topics related to the understanding of human evolution, genetics and reproduction has been covered in this chapter, such as anatomy of bipedalism, behavioral modernity, evolution of human intelligence, human DNA, RNA, etc.

Human Evolution

Human evolution is the process by which human beings developed on Earth from now-extinct primates. Viewed zoologically, we humans are *Homo sapiens*, a culture-bearing, upright-walking species that lives on the ground and very likely first evolved in Africa about 315,000 years ago. We are now the only living members of what many zoologists refer to as the human tribe, Hominini, but there is abundant fossil evidence to indicate that we were preceded for millions of years by other hominins, such as *Australopithecus*, and that our species also lived for a time contemporaneously with at least one other member of our genus, *Homo neanderthalensis* (the Neanderthals). In addition, we and our predecessors have always shared the Earth with other apelike primates, from the modern-day gorilla to the long-extinct *Dryopithecus*. That we and the extinct hominins are somehow related and that we and the apes, both living and extinct, are also somehow related is accepted by anthropologists and biologists everywhere. Yet the exact nature of our evolutionary relationships has been the subject of debate and investigation since the great British naturalist Charles Darwin published his monumental books *On the Origin of Species* (1859) and *The Descent of Man* (1871).

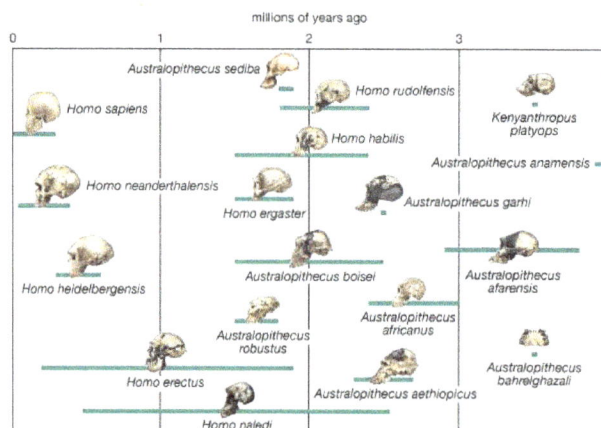

Possible pathways in the evolution of the human lineage

The primary resource for detailing the path of human evolution will always be fossil specimens. Certainly, the trove of fossils from Africa and Eurasia indicates that, unlike today, more than one species of our family has lived at the same time for most of human history. The nature of specific fossil specimens and species can be accurately described, as can the location where they were found and the period of time when they lived; but questions of how species lived and why they might have either died out or evolved into other species can only be addressed by formulating scenarios, albeit scientifically informed ones. These scenarios are based on contextual information gleaned from localities where the fossils were collected. In devising such scenarios and filling in the human family bush, researchers must consult a large and diverse array of fossils, and they must also employ refined excavation methods and records, geochemical dating techniques, and data from other specialized fields such as genetics, ecology and paleoecology, and ethology (animal behaviour)—in short, all the tools of the multidisciplinary science of paleoanthropology.

Anatomy of Bipedalism

Bipedalism is not unique to humans, though our particular form of it is. Whereas most other mammalian bipeds hop or waddle, we stride. *Homo sapiens* is the only mammal that is adaptedexclusively to bipedal striding. Unlike most other mammalian orders, the primates have hind-limb-dominated locomotion. Accordingly, human bipedalism is a natural development from the basic arboreal primate body plan, in which the hind limbs are used to move about and sitting upright is common during feeding and rest.

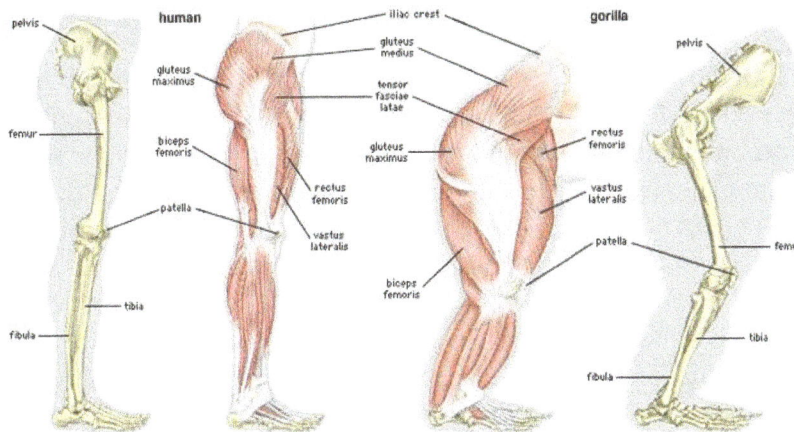

Skeletal and muscular structures of a human leg (left) and a gorilla leg (right)

The initial changes toward an upright posture were probably related more to standing, reaching, and squatting than to extended periods of walking and running. Human beings stand with fully extended hip and knee joints, such that the thighbones are aligned with their respective leg bones to form continuous vertical columns. To walk, one simply tilts forward slightly and then keeps up with the displaced centre of mass, which is located within the pelvis. The large muscle masses of the human lower limbs power our locomotion and enable a person to rise from squatting and sitting postures. Body mass is transferred through the pelvis, thighs, and legs to the heels, balls of the feet, and toes. Remarkably little muscular effort is expended to stand in place. Indeed, our large buttock, anterior thigh, and calf muscles are virtually unused when we stand still. Instead of muscular contraction, the human bipedal stance depends more on the way in which joints are constructed and on strategically located ligaments that hold the joints in position. Fortunately for paleoanthropologists, some bones show dramatic signs of how a given hominin carried itself, and the adaptation to obligate terrestrial bipedalism led to notable anatomic differences between hominins and great apes. These differences are readily identified in fossils, particularly those of the pelvis and lower limbs.

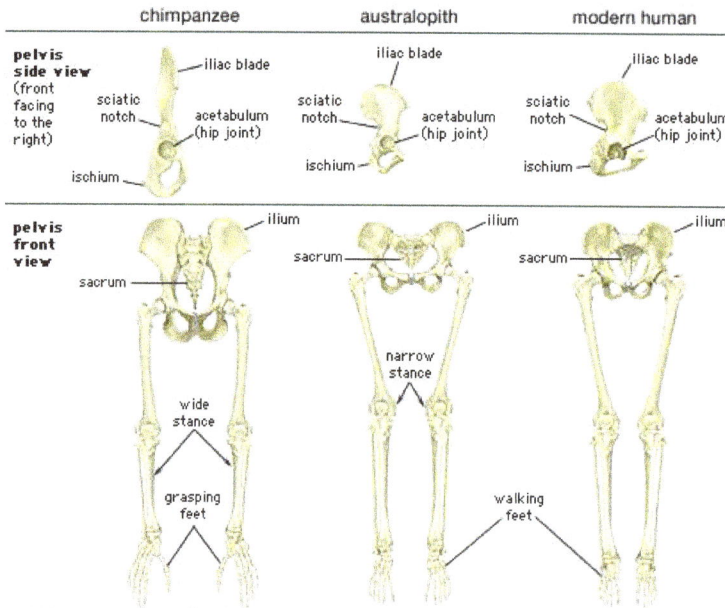

Comparison of the pelvis and lower limbs of a chimpanzee, an australopith, and a modern human

Although we are bipedal, our pelvis is oriented like that of quadrupedal primates. The early bipedal hominins assumed erect trunk posture by bending the spine upward, particularly in the lower back (lumbar region). In order to transfer full upper-body mass to the lower limbs and to reposition muscles so that one could walk without assistance from the upper limbs and without wobbling from side to side, changes were required in the pelvis—particularly in the ilia (the large, blade-shaped bones on either side), the ischia (protuberances on which body rests when sitting), and the sacrum (a wedge-shaped bone formed by the fusing of vertebrae). Hominin hip bones have short ilia with large areas that articulate with a short, broad sacrum. Conversely, great-ape hip bones have long ilia with small sacral articular areas, and sacra of the great apes are long and narrow. The human pelvis is unique among primates in having the ilia curved forward so that the inner surfaces face one another instead of being aligned sideways, as in apes and other quadrupeds. Curved ilia situate some of the gluteal muscles on the side of the hip joint, where they steady the pelvis as the foot swings forward during a step. This special mechanism allows us to walk smoothly, with only slight oscillations of the pelvis and without gross side-to-side motions of the upper body. Humans have short ischia (and long lower limbs), facilitating speedy actions of the hamstring muscles, which extend the thigh at the hip joint, while great apes have long ischia (and short hind limbs), which give them powerful hip extension for climbing up trees. Characteristically, a human thighbone is long and has a very large, globular head and a short, round neck; at the knee a prominent lateral ridge buttresses the groove in which the kneecap lies. The femurs are farther apart at the hips than at the knees and slant toward the midline to keep the knees close together. This angle allows anthropologists to diagnose bipedalism even if the fossil is only the knee end of a femur. The femurs of quadrupedal great apes, on the other hand, do not converge toward the knees, and the femoral shafts lack telltale angling.

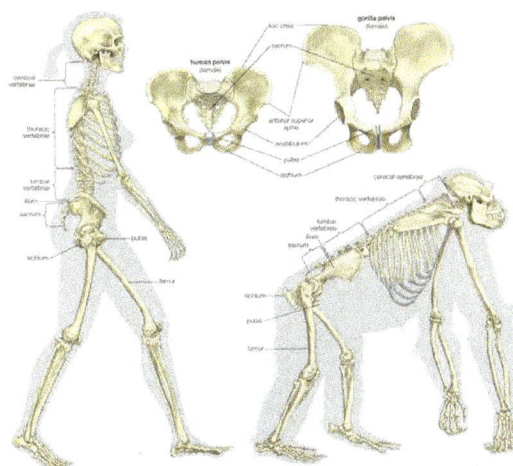

The skeletal structure of a human being (left) and of a gorilla (right)

Several differences allow the human being to walk erect on two legs with a striding gait rather than move in a knuckle-walking fashion like the gorilla. In the pelvis these differences include shorter ischia, a broader sacrum, and broader, curved-in ilia with a lower iliac crest. In the legs the femurs (thigh-bones) are relatively long and are set farther apart at the hips than they are at the knees.

Human feet are distinct from those of apes and monkeys. This is not surprising, since in humans the feet must support and propel the entire body on their own instead of sharing the load with the forelimbs. In humans the heel is very robust, and the great toe is permanently aligned with the four diminutive lateral toes. Unlike other primate feet, which have a mobile midfoot, the human foot possesses (if not requires) a stable arch to give it strength. Accordingly, human footprints are unique and are readily distinguished from those of other animals.

Fossil Evidence

A trail of footprints probably left by *Australopithecus afarensis*
individuals some 3.5 million years ago, at Laetoli, northern Tanzania

By 3.5 million years ago at least one hominin species, *A. afarensis*, was an adept walker. In addition to anatomic evidence from this time, there is also a 27.5-metre (90-foot) trackway produced by three individuals who walked at a leisurely pace on moist volcanic ash at Laetoli in northern Tanzania.

In all observable features of foot shape and walking pattern, they are astonishingly similar to those of habitually barefoot people who live in the tropics today. Nevertheless, although the feet of the Laetoli hominins appear to be strikingly human, one should not assume that other parts of their bodies were as similar to ours.

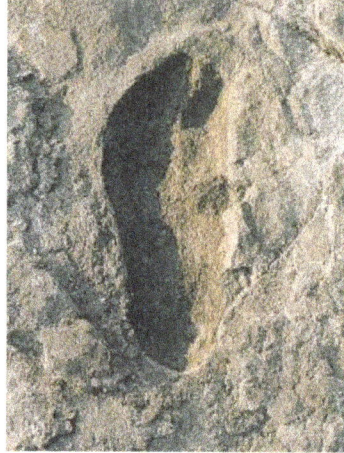

A single footprint of *Australopithecus afarensis* (top), left some 3.5 million years ago at Laetoli, Tanzania, shows a striking similarity to a single footprint of a habitually barefoot modern human being from Peru (bottom)

The fragmentary femoral remains found in Kenya of six-million-year-old *Orrorin tugenensis* indicate to some experts that they too were bipeds. *Ardipithecus ramidus* (5.8–4.4 mya), a primate from Aramis, central Ethiopia, was also bipedal. In this case the evidence comes from the foramen magnum, the hole in the skull through which the spinal cord enters. In *Ardipithecus* this opening is similar to ours in being located centrally under the skull instead of at the rear of it. A rear-facing foramen magnum indicates a stooped posture, whereas a downward-facing hole positions the skull atop the spinal column. Other characteristics indicative of bipedalism in *Ardipithecus* include an increased tarsal region in each foot and a pelvic structure with muscle-to-bone attachment sites comparable to later, bipedal hominins. In addition, the leg bone of *Australopithecus anamensis* from northern Kenya (4.2–3.9 mya) attests to its bipedalism.

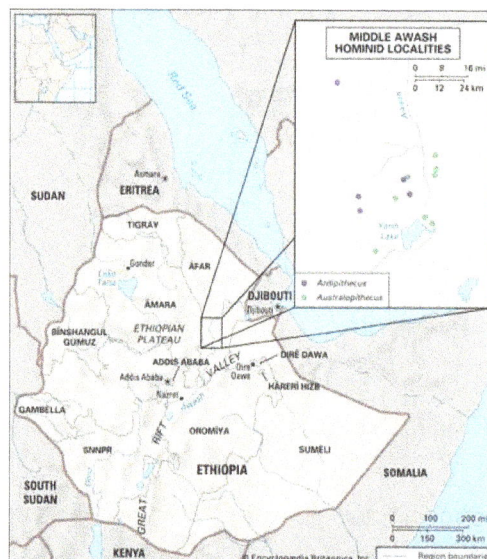

Hominin fossil sites of the Awash River basin, Ethiopia

All hominins living at the time of the Laetoli track makers were probably obligate bipeds when on the ground, but some of them (including some younger species) exhibit features that argue for regular arboreal climbing, probably for food, rest, nightly lodging, and predator avoidance. Hadar, in northern Ethiopia, has yielded a trove of remains of *A. afarensis* (3.8–2.9 mya). They include many parts of the locomotor skeleton that reveal a bipedal habit: short ilia, a wide and stout sacrum, and femoral angling, among other features. At the same time, the curved fingers and toes, laterally flared ilia, and short femurs with long upper limbs, as well as the configuration of its rib cage, indicate that they could readily climb and maneuver in trees. *A. bahrelghazali* (3.5–3.0 mya) of central Chad and *Kenyanthropus platyops* (3.5 mya) from northern Kenya are represented solely by teeth and by skull and jaw fragments from which positional behaviour cannot be inferred.

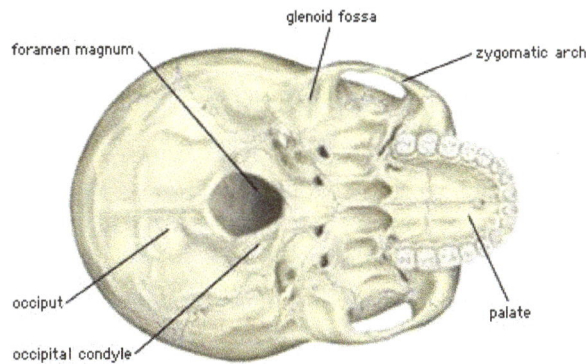

View of the base of the human skull, showing the central location of the foramen magnum

Parts of the locomotor skeletons of later hominins such as *A. africanus* (3.3–2.4 mya) and *Paranthropus robustus* (1.8–1.5 mya) of South Africa do not differ markedly from those of *A. afarensis*. The locomotor skeleton of eastern African *P. boisei* (2.2–1.3 mya) is poorly known, but there is no reason to assume that it was different from other *Paranthropus* species. Bouri, a 2.5-million-year-old site in central Ethiopia, yielded arm and leg bones that are contemporaneous with craniodental remains of *A. garhi*. The femur is elongated relative to the humerus, as in *Homo sapiens*, but, unlike the human forearm, that of the fossil specimen is relatively long. Thus, by 2.5 mya at least one hominin species had developed the long femurs of striding bipeds, though it retained long forearms like arboreally active *Australopithecus* and *Paranthropus*.

Homo habilis (2.0–1.5 mya), best known from Olduvai Gorge, Tanzania, exhibits small teeth and a large brain, but it has long upper limbs (especially the forearms), short femurs, curved finger bones, and other chimpanzee-like traits that indicate a mélange of arboreal and terrestrial adaptations. Because of these similarities, some investigators classify *H. habilis* as *A. habilis*.

The pelvis of *H. heidelbergensis* (600,000–200,000 years ago, or 600–200 kya) and that of Neanderthals (200–30 kya) are distinct from the pelvis of *Homo sapiens* in some features that recall those of *Australopithecus*. The pelvis is broad, with ilia flaring out to the side. The femoral necks are also relatively long. These features are related to stabilizing the pelvis in stocky bipedal hominins. The pelvises of both *H. heidelbergensis* and Neanderthals could accommodate a wider birth canal. This feature is important because they may have had notably larger brains (about 1,200 grams [2.65 pounds] and 1,400 grams [3.09 pounds], respectively) than earlier hominins did—a trait that is reflected in the size of the fetal skull.

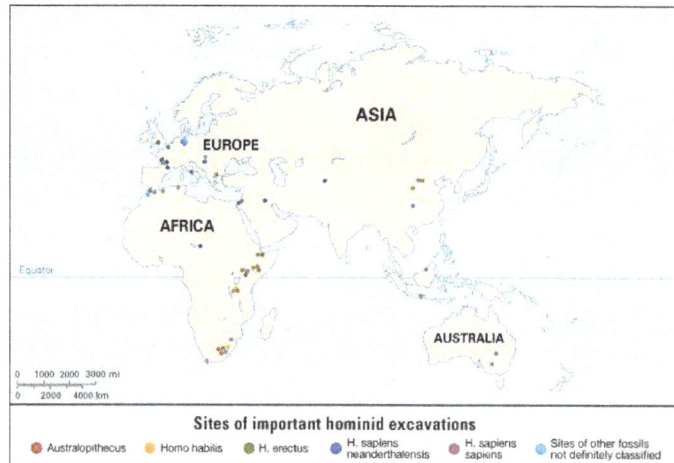

The earliest fossils of human ancestors have been found in Africa

Regrettably, development of foot structure in early *Homo*—i.e., between *A. afarensis* and Neanderthals—is virtually undocumented by skeletal evidence. The oldest footprints indicative of contemporary foot function, however, have been found in Ileret, Kenya. These prints have been dated at 1.51 to 1.53 mya, and their size and depth suggest that they were made by *H. ergaster* or *H. erectus*. Therefore, it is safe to assume that by about 1.53 mya the uniquely human locomotor and associated cooling systems were basically established. Subsequent alterations in pelvic shape may be related to the passage of larger-brained babies through the birth canal.

Refinements in Hand Structure

Primates are hand-to-mouth feeders that pluck and catch items selectively by hand before ingesting them. Without tools, emergent hominins would have relied on the versatility and strength of their hands to collect food and on their teeth and jaws alone to process it. Unless they used tools to fashion carrying devices such as bags from animal skins, they would have needed a reliable source of water nearby, and they would also have been limited in the types and number of objects that they could transport through their range. In addition to transporting objects and water, there is the more obvious utility of animal skins in protecting against night chills, rain, and strong sunshine.

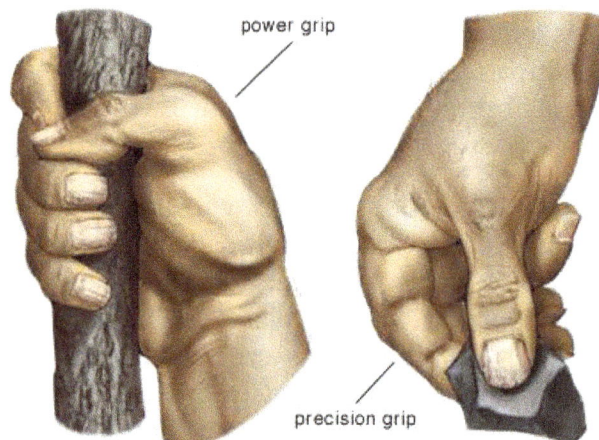

A fully opposable thumb gives the human hand its unique power grip (left) and precision grip (right)

Sharp-edged stones, even small flakes, would be a boon to early hominins who learned how to select and make them for cutting hides, meat, sticks, and other plant material. Stones also would assist in pounding open hard-shelled fruits and nuts, bones for marrow, and skulls for brains. There may have been a span when early hominins used naturally occurring stones and other objects as tools and weapons, much as some wild chimpanzees do today.

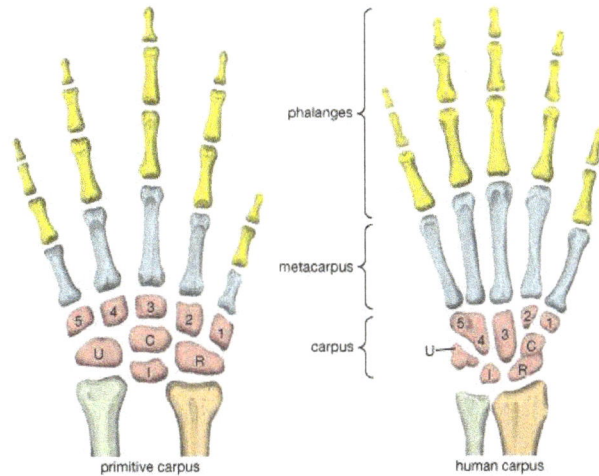

carpal bonesCarpal bones of a primitive humanoid and a human

Before hominins controlled fire and either built sturdy shelters on the ground or effectively defended caves and rock shelters, they may have constructed platforms in trees for daily activities as well as night lodging. Raw materials, stone hammers, cutting tools, and sticks and stones for defense could be stored in the trees to be used repeatedly. Handheld rocks, clubs, and long stabbing sticks, spears, or other missiles would constitute a formidable defense, especially if employed from the vantage of a tree platform.

By about 2.6 million years ago, some hominins were making and using simple stone artifacts in eastern Africa. A likely candidate for this practice is *H. habilis*, though its contemporaries *P. boisei* and *A. garhi* cannot be overruled for this distinction. Indeed, at Bouri, Ethiopia, mammalian bones that were cut and pounded by stone tools occur in 2.5-million-year-old sediments contemporaneous with those yielding *A. garhi*.

Because the earliest stone artifacts were of such simple construction and because chimpanzees, orangutans, and capuchin monkeys today can employ stones, stems, vines, and sticks to extract nutritious morsels from protective covers, one need not expect that early hominin toolmakers displayed modern hand structure and exquisite motor control. Nonetheless, the unique structure of the human hand is readily explained by a substantial history of producing and using increasingly complex tool kits and other artifacts.

The features of human hands are easily distinguishable from those of the great apes, and they underpin our refined manipulatory abilities. The most complex adaptations of the human hand involve the thumb, wherein a unique, fully independent muscle (the flexor pollicis longus) gives this digit remarkable strength in pinch and power grips. The fingertips are broad and equipped with highly sensitive pads of skin. The proportional lengths of the thumb and other fingers give us an opposable thumb with precise, firm contact between its tip and the ends of each of the other fingers. A special saddle joint and associated ligaments at the base of the thumb facilitate refined rotation. Special configurations of joints at the bases of the fifth, fourth, and second fingers facilitate tip-to-tip precision grips with the thumb. Asymmetry of the heads of the second and fifth palm bones induces rotation of the articulated fingers during opposition with the thumb. Finally, numerous modifications of the small muscles in the hand are associated with fine control of the thumb and fingers.

Australopithecus afarensis is the earliest hominin species for which there are sufficient fossil hand bones to assess manipulatory capabilities. They were capable of gripping sticks and stones firmly for vigorous pounding and throwing, but they lacked a fully developed human power grip that would allow cylindrical objects to be held between the partly flexed fingers and the palm, with counterpressure being applied by the thumb. There are insufficient specimens to assess fine manipulation in *Australopithecus*, but there is no reason to believe that they were less capable than modern chimpanzees. Chimpanzees and other apes have remarkable precision of grip, even though the tapered thumb tip must be pressed against the side of the index finger and cannot be apposed securely to any of the fingertips.

Hand bones assigned to a 1.8-million-year-old specimen of *H. habilis* from Olduvai Gorge in northern Tanzania represent an advance over those of *A. afarensis* in features related to tool use. Tools similar to those found at Olduvai are found associated with *H. habilis* from other parts of eastern Africa as well. The tips of its thumb and fingers were flat, and there is evidence for a strong flexor pollicis longus muscle and a saddle joint at the base of the thumb. Hand bones arguably assigned to *P. robustus* or *Homo* from Swartkrans, South Africa, confirm that by about 1.8 mya one or more hominin species had highly developed thumbs and flat fingertips.

Hominin hand bones from 2.8–2.5-million-year-old cave deposits at Sterkfontein, South Africa, may be evidence that the hands of *A. africanus* were somewhat more advanced for stone tool use, but no artifact has been found in association with them. Younger Sterkfontein deposits (2.0–1.5 mya) contain stone artifacts and remains of a *Homo* species.

Because of an absence of fossils, it is not possible to track certain refinements in hand structure that must have evolved in conjunction with innovations in tool manufacture and use during the heydays of *H. rudolfensis*, *H. ergaster* (1.9–1.5 mya), and *H. erectus* (1.7–0.2 mya), as well as *H. antecessor* (1.0–0.8 mya) and *H. heidelbergensis* (600–200 kya). Only prehistoric and modern *Homo sapiens* and *H. neanderthalensis* are fully represented by hand skeletons.

Increasing Brain Size

Because more complete fossil heads than hands are available, it is easier to model increased brain size in parallel with the rich record of artifacts from the Paleolithic Period (c. 2,500,000 to 10,000 years ago), popularly known as the Old Stone Age. The Paleolithic preceded the Middle Stone Age, or Mesolithic Period; this nomenclature sometimes causes confusion, as the Paleolithic itself is divided into Early, Middle, and Late (or Upper) periods. Hominin brain expansion tracks so closely with refinements in tool technology that some scholars ignore other factors that may have contributed to the brain's increasing size, such as social complexity, foraging strategies, symbolic communication, and capabilities for other culture-mediated behaviours that left no or few archaeological traces.

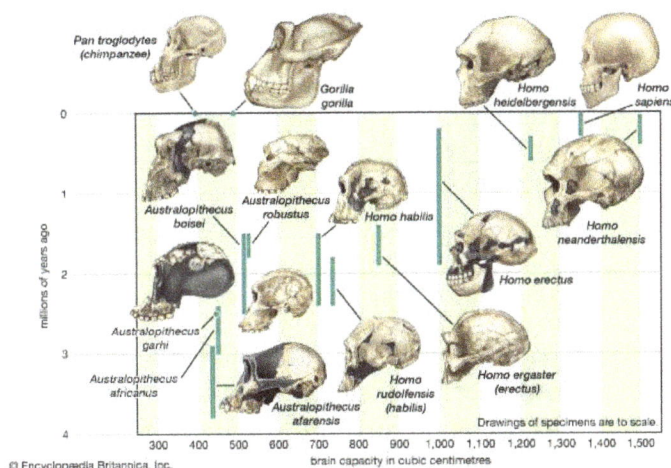

The increase in hominin cranial capacity over time

Throughout human evolution, the brain has continued to expand. Estimated average brain masses of *A. afarensis* (435 grams [0.96 pound]), *A. garhi* (445 grams [0.98 pound]), *A. africanus* (450 grams [0.99 pound]), *P. boisei* (515 grams [1.13 pounds]), and *P. robustus* (525 grams [1.16 pounds]) are close to those of chimpanzees (395 grams [0.87 pound]) and gorillas (490 grams [1.08 pounds]). Average brain mass of *Homo sapiens* is 1,350 grams (2.97 pounds). The increase appears to have begun with *H. habilis* (600 grams [1.32 pounds]), which is also notable for having a small body. The trend in brain enlargement continued in Africa with larger-bodied *H. rudolfensis* (735 grams [1.62 pounds]) and especially *H. ergaster* (850 grams [1.87 pounds]).

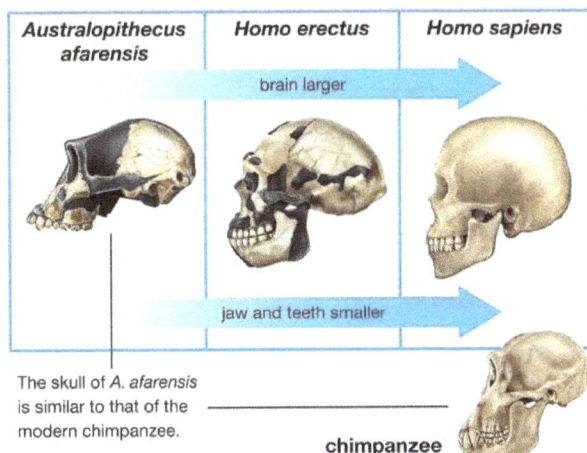

hominin cranial capacityThe evolution of relative cranial capacity and dentition patterns in selected hominins

One must be extremely cautious about ascribing greater cognitive capabilities, however. Relative to estimated body mass, *H. habilis* is actually "brainier" than *H. rudolfensis* and *H. ergaster*. A similar interpretive challenge is presented by Neanderthals versus modern humans. Neanderthals had larger brains than earlier *Homo* species, indeed rivaling those of modern humans. Relative to body mass, however, Neanderthals are less brainy than anatomically modern humans. Relative brain size of *Homo* did not change from 1.8 to 0.6 mya. After about 600 kya it increased until about 35,000 years ago, when it began to decrease. Worldwide, average body size also decreased in *Homo sapiens* from 35,000 years ago until very recently, when economically advanced peoples began to grow larger while less-privileged peoples did not.

hominin	number of fossil examples	average capacity of the braincase (cc)
Australopithecus	6	440
Paranthropus	4	519
Homo habilis	4	640
Javanese Homo erectus (Trinil and Sangiran)	6	930
Chinese Homo erectus (Peking man)	7	1,029
Homo sapiens	7	1,350
Average capacity of the braincase in fossil hominins		

Overall, there were periods of stagnation and elaboration in stone tool technology during the Paleolithic, but, because of variations over time and between locations as well as the possibility that plant materials were used instead of stone, it is impossible to link brain size with technological complexity and fully human cognitive capabilities. Moreover, in many instances it is impossible to identify assuredly the hominin species that commanded a Paleolithic industry, even when there are associated skeletal remains at the site.

The unreliability of brain size to predict cognitive competence and ability to survive in challenging environments is underscored by the discovery of a distinctive human sample, dubbed *Homo floresiensis*, in a limestone cave on Flores Island, Indonesia, in 2004. The diminutive *H. floresiensis* had brains comparable in mass to those of chimpanzees and small australopiths, yet they produced a stone tool industry comparable to that of Early Pleistocene hominins and survived among giant rats, dwarf elephants, and Komodo dragons from at least 38 kya to about 18 kya. If they are indeed a distinct species, they constitute yet another archaic human (in addition to *H. neanderthalensis* and perhaps *H. erectus*) that lived contemporaneously with modern humans during the Late Pleistocene.

Emergence of Homo Sapiens

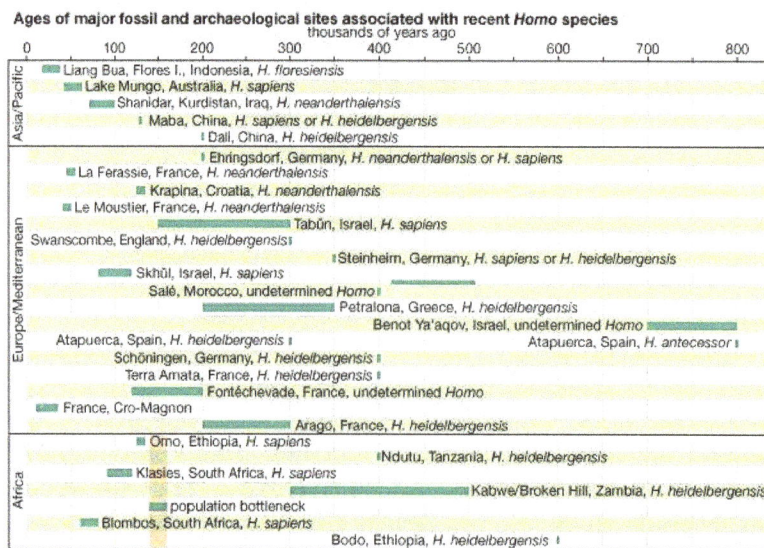

Ages of major fossil and archaeological sites associated with recent *Homo* species

The relationships among *Australopithecus*, *K. platyops*, *Paranthropus*, and the direct ancestors of *Homo* are unknown. Because of its early date and geographic location, *A. anamensis* may be the common ancestor of *A. afarensis*, *A. garhi*, *K. platyops*, and perhaps the Laetoli Pliocene hominins of eastern Africa, *A. bahrelghazali* of central Africa, and *A. africanus* of southern Africa. *A. afarensis* in turn may be ancestral to *P. aethiopicus*, which begat *P. boisei* in eastern Africa and *P. robustus* in southern Africa.

Factors indicating *H. rudolfensis* as ancestral to later species of *Homo* are its absolute brain size, large body, and lower limb morphology. These features clearly foreshadow younger species of *Homo* in Africa and Eurasia. However, a mandible discovered in the Ledi-Geraru area of the Awash River valley in 2013 may point toward a different ancestor—one that clearly belongs to the genus *Homo*. The mandible provides evidence that dental features associated with later *Homo* (such as smaller teeth and a much-reduced chin) appeared as early as 2.8 million years ago, well in advance of the advent of *H. rudolfensis*. While some paleontologists have been quick to associate the specimen with *H. habilis*, others are considering the possibility that it belongs to a new species of *Homo*.

Our ancestry becomes no clearer as the candidates are narrowed to *Homo* species exclusively. Among paleoanthropologists who accept it as a species distinct from *H. erectus*, *H. ergaster* is most often proposed as the ancestor of *Homo* species of the Pleistocene Epoch. *H. heidelbergensis* may have arisen from *H. ergaster*, *H. erectus*, or *H. antecessor*, and any or none of them could have been ancestors of *H. neanderthalensis* and *Homo sapiens*. Neanderthal populations, particularly as represented by specimens from western Europe, probably were not ancestral to modern humans.

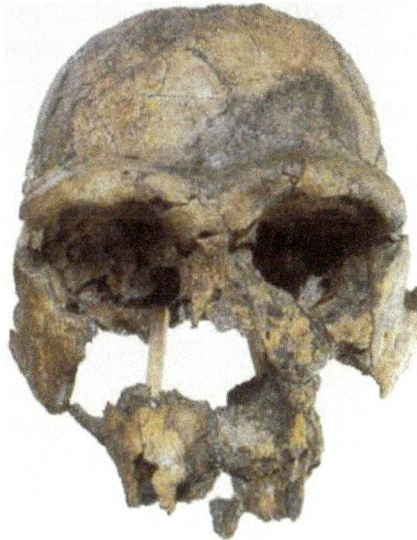

Replica of KNM-ER 3733, a 1.75-million-year-old *Homo erectus* skull found in 1975 at Koobi Fora, Kenya

Theorists use fossil remains, genetic traits of modern people around the world, and archaeological and anatomical indicators of cognitive, linguistic, and technological capabilities to support their models of recent human evolution, but no single theory provides definitive resolution of how *Homo sapiens* came to be. The limitations of empirical evidence confound efforts to discern whether distinctive features and lineages developed gradually or over periods of stasis punctuated by rapid change (a theory known as punctuated equilibrium). There are claims for about 20 fossil hominin-species over the course of the last six million years, but they are assessed on a case-by-case basis. For example, it appears that Neanderthals (*H. neanderthalensis*) were a dead end for two ancestral species (*H. antecessor* and *H. heidelbergensis*) that changed gradually in Europe from about 700 kya to 30 kya. *Homo sapiens* may have evolved similarly through a series of species represented by African specimens, but other theorists envision a dramatic shift in cognitive capacity and behaviour that qualifies instead as a punctuational change. This change would have occurred in one small African population and would have been followed by a long period of stasis that continues to the present. Such a scenario is not unprecedented, as *A. afarensis* was a capable biped that appears to have emerged suddenly and persisted for nearly one million years.

There are four basic models that purport to explain the evolution of *Homo sapiens* between about 200 and 30 kya. At one extreme is multiregional evolution, or the regional continuity model. At the other is the African replacement, or "out of Africa," model. Intermediate are the African hybridization-and-replacement model and the assimilation model. All but the multiregional model maintain that *Homo sapiens* evolved solely in Africa and then deployed to Eurasia and eventually the Americas and Oceania. Both of the replacement models argue that anatomically modern emigrants replaced resident Eurasian and Australasian species of *Homo sapiens* with little or no hybridization. The hybridization-and-replacement model proposes some interbreeding with archaic indigenouspopulations but with relatively minor effects. Assimilation maintains continuity between archaic and modern humans, most notably in some areas of Eurasia, where gene flow and local selective factors would also produce morphological changes. In this model, unity of the species was maintained by periodic interbreeding across wide areas. Multiregionalists reject the idea that *Homo sapiens* evolved uniquely in Africa. Instead, they advocate that discrete archaic populations of *Homo* evolved locally in Africa, Asia, and Europe. Throughout their tenures, both the archaic and descendant populations interbred with contemporaries from other areas.

The African replacement model has gained the widest acceptance owing mainly to genetic data (particularly mitochondrial DNA) from existing populations. This model is consistent with the realization that modern humans cannot be classified into subspecies or races, and it recognizes that all populations of present-day humans share the same potential.

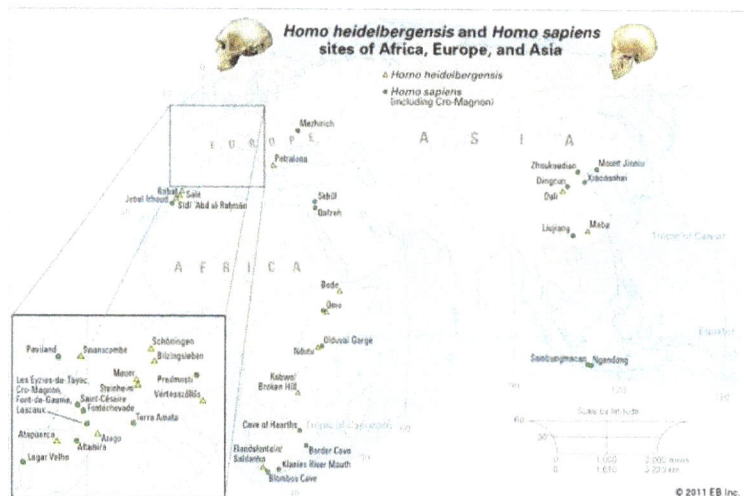

Sites of Homo heidelbergensis and Homo sapiens remainsSites of Homo heidelbergensis and Homo sapiens remains in Africa, Europe, and Asia

Such a tangled line of descent is not surprising given the nomadic lifestyles enabled by bipedalism. There appear to have been successive migrations of hominin species out of Africa, with evolution of new species in Eurasia and occasional migrations back into Africa. For instance, *H. ergaster* may have been the first hominin to reach Eurasia. Some of its descendants could have moved quickly to East and Southeast Asia, where they begat *H. erectus*. Others may have evolved into *H. heidelbergensis*, which populated Europe sparsely and then returned to Africa.

Some paleoanthropologists claim that *H. antecessor*, found in 800,000-year-old cave deposits at Gran Dolina, Sierra de Atapuerca, Spain, was a direct ancestor of *H. neanderthalensis* via *H. heidelbergensis*, which is represented by 300,000-year-old specimens from Sima de los Huesos in the Sierra de Atapuerca. Further, they propose that *H. antecessor*, from million-year-old deposits in Eritrea, is a direct ancestor of *Homo sapiens* in Africa.

Neanderthals probably evolved in Europe at least partially in response to cold climatic conditions and then migrated to western Asia, where they may have encountered *Homo sapiens* in the Levant. There is no skeletal evidence that they reached the African continent or moved much farther east than Uzbekistan in Central Asia. Features of Neanderthals that argue for adaptation to seasonally frigid biomes include stocky torsos, short limbs (particularly the forearms and legs), and distinctive facial structure. The middle of the face protrudes, the teeth are set forward, the enlarged cheekbones sweep backward, and the nasal passages are voluminous. If Neanderthals wore animal furs and other insulating materials on their heads and bodies while keeping vigorously active in frigid weather, the large nasal chamber would help to cool the blood and prevent overheating the brain, while clothing would reduce the risk of frostbite. The nasal chamber might also conserve moisture during exhalation.

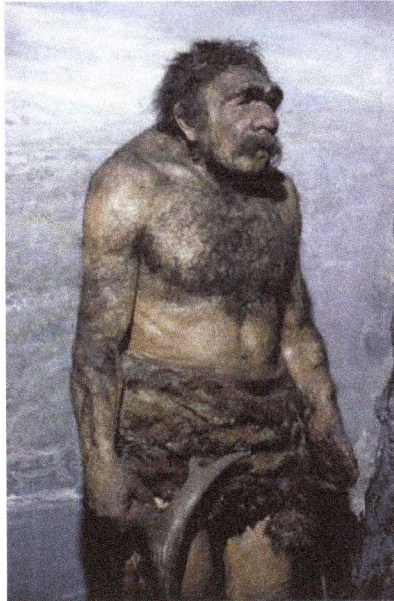

Reconstructed model of a male Neanderthal (*Homo neanderthalensis*)

Fossil specimens obtained from the Omo site in Ethiopia (which have been dated to 195 kya) indicate that anatomically modern *Homo sapiens* were present sometime around 200 kya in eastern Africa. The oldest known remains, however, appear at the Jebel Irhoud site in Morocco and date to 315 kya. This evidence suggests that the species might not have emerged in eastern Africa or that it was not confined to the region. Molecular genetic data suggest that early *Homo sapiens* passed through a population bottleneck—that is, a period when they were rare creatures—before rapidly spreading throughout the Old World. *Homo sapiens* migrated to southern China between 120 kya and 80 kya and Europe about 45–43 kya. They replaced indigenous hominin species in Eurasia, and then, as sea levels dropped during glacial periods, adventurous individuals went to sea in watercraft, populating Australia about 50 kya and oceanic islands during the most recent 3,000 years. Most evidence points to *Homo sapiens* migrating to the Americas about 14–13.3 kya; however, some evidence suggests that this migration may have taken place up to 15,000 years earlier.

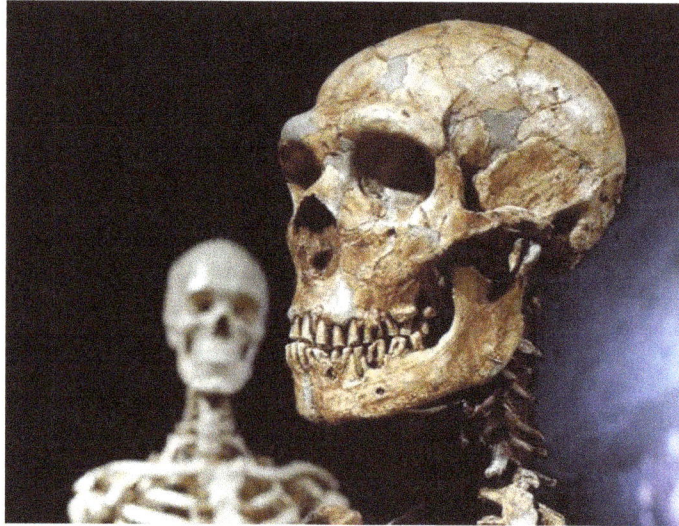

Neanderthal Replica skull of a Neanderthal (*Homo neanderthalensis*), with a replica skeleton of a modern human (*Homo sapiens*) in the background

Some of the extensive variation in bodily proportions, external features, and blood chemistry of modern peoples may reflect adjustments to biomes over geologically short time spans. However, molecular genetic studies show that genomic differences between even far-flung peoples are minuscule compared with variations within each local population. Accordingly, for modern *Homo sapiens*, race is a mere cultural construct with no biological basis.

Evolution of Human Intelligence

The human brain is a fuel hog, and that, it turns out, is key to how our intelligence evolved. It has long been believed that the evolution of human intelligence was simply related to increasing brain size, but a team of researchers from South Africa and Australia have overturned that assumption.

By calculating how blood supply to the brains of human ancestors changed over time, the researchers were able to show that the human brain not only evolved to become larger, but also more blood-thirsty. And the need for blood outpaced the volume increase of the brain itself.

"Brain size has increased about 350% over human evolution, but we found that blood flow to the brain increased an amazing 600%," project leader Roger Seymour, from the University of Adelaide, said in a statement. "We believe this is possibly related to the brain's need to satisfy increasingly energetic connections between nerve cells that allowed the evolution of complex thinking and learning."

So the more nerve cells and connections our brains developed, the more metabolic activity increased. And that brought a higher demand for the oxygen and nutrients that blood provides.

Two holes at the base of the human skull allow arteries to shuttle blood from the heart to the brain. The diameters of the holes correspond to the size and blood-carrying capacity of the arteries. By tracking the changes in the size of those holes in human ancestors, the researchers were able to track the evolution of human intelligence.

"The intensity of brain activity was, before now, believed to have been taken to the grave with our ancestors," said co-author Edward Snelling from the University of the Witwatersrand in South Africa.

In a study published last year, the same researchers reported that the intelligence of animal species could be predicted by measuring the size of the skull openings.

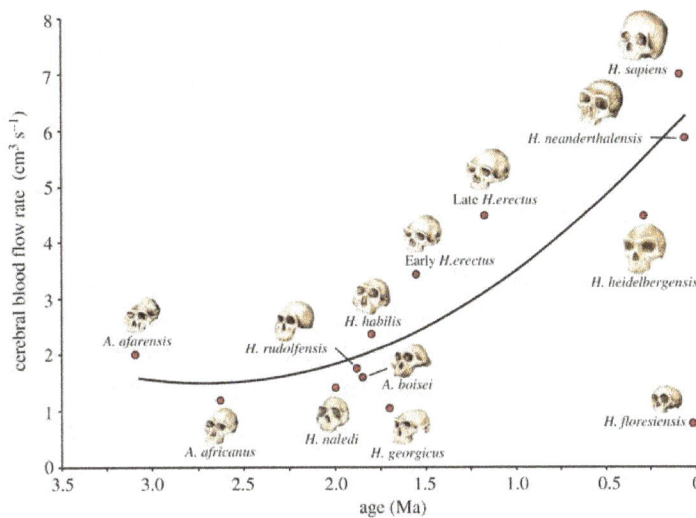

Cerebral blood flow rate in relation to estimated geological age in 12 hominin species

Human Intelligence Evolution Models

Social brain hypothesis

The social brain hypothesis was proposed by British anthropologist Robin Dunbar, who argues that human intelligence did not evolve primarily as a means to solve ecological problems, but rather as a means of surviving and reproducing in large and complex social groups. Some of the behaviors associated with living in large groups include reciprocal altruism, deception and coalition formation. These group dynamics relate to Theory of Mind or the ability to understand the thoughts and emotions of others, though Dunbar himself admits in the same book that it is not the flocking itself that causes intelligence to evolve (as shown by ruminants).

Dunbar argues that when the size of a social group increases, the number of different relationships in the group may increase by orders of magnitude. Chimpanzees live in groups of about 50 individuals whereas humans typically have a social circle of about 150 people, which is also the typical size of social communities in small societies and personal social networks; this number is now referred to as Dunbar's number. In addition, there is evidence to suggest that the success of groups is dependent on their size at foundation, with groupings of around 150 being particularly successful, potentially reflecting the fact that communities of this size strike a balance between the minimum size of effective functionality and the maximum size for creating a sense of commitment to the community. According to the social brain hypothesis, when hominids started living in large groups, selection favored greater intelligence. As evidence, Dunbar cites a relationship between neocortex size and group size of various mammals.

Criticism

Phylogenetic studies of brain sizes in primates show that while diet predicts primate brain size, sociality does not predict brain size when corrections are made for cases in which diet affects both brain size and sociality. The exceptions to the predictions of the social intelligence hypothesis, which that hypothesis has no predictive model for, are successfully predicted by diets that are either nutritious but scarce or abundant but poor in nutrients.

Meerkats have far more social relationships than their small brain capacity would suggest. Another hypothesis is that it is actually intelligence that causes social relationships to become more complex, because intelligent individuals are more difficult to learn to know.

There are also studies that show that Dunbar's number is not the upper limit of the number of social relationships in humans either.

The hypothesis that it is brain capacity that sets the upper limit for the number of social relationships is also contradicted by computer simulations that show simple unintelligent reactions to be sufficient to emulate "ape politics" and by the fact that some social insects such as the paper wasp do have hierarchies in which each individual has its place (as opposed to herding without social structure) and maintains their hierarchies in groups of approximately 80 individuals with their brains smaller than that of any mammal.

Social Exchange Theory

Other studies suggest that social exchange between individuals is a vital adaptation to the human brain, going as far to say that the human mind could be equipped with a neurocognitive system specialized for reasoning about social change. Social Exchange is a vital adaptation that evolved in social species and has become exceptionally specialized in humans.This adaption will develop by natural selection when two parties can make themselves better off than they were before by exchanging things one party values less for things the other party values for more. However, selection will only pressure social exchange when both parties are receiving mutual benefits from their relative situation; if one party cheats the other by receiving a benefit while the other is harmed, then selection will stop. Consequently, the existence of cheaters—those who fail to deliver fair benefits—threatens the evolution of exchange. Using evolutionary game theory, it has been shown that adaptations for social exchange can be favored and stably maintained by natural selection, but only if they include design features that enable them to detect cheaters, and cause them to channel future exchanges to reciprocators and away from cheaters. Thus, humans use social contracts to lay the benefits and losses each party will be receiving (if you accept benefit B from me, then you must satisfy my requirement R). Humans have evolved an advanced cheater detection system, equipped with proprietary problem-solving strategies that evolved to match the recurrent features of their corresponding problem domains. Not only do humans need to determine that the contract was violated, but also if the violation was intentionally done. Therefore, systems are specialized to detect contract violations that imply intentional cheating.

One problem with the hypothesis that specific punishment for intentional deception could co-evolve with intelligence is the fact that selective punishment of individuals with certain characteristics selects against the characteristics in question. For example, if only individuals capable of remembering what they had agreed to were punished for breaking agreements, evolution would have selected against the ability to remember what one had agreed to.

Sexual Selection

This model, which invokes sexual selection, is proposed by Geoffrey Miller who argues that human intelligence is unnecessarily sophisticated for the needs of hunter-gatherers to survive. He argues that the manifestations of intelligence such as language, music and art did not evolve because of their utilitarian value to the survival of ancient hominids. Rather, intelligence may have been a fitness indicator. Hominids would have been chosen for greater intelligence as an indicator of healthy genes and a Fisherian runawaypositive feedback loop of sexual selection would have led to the evolution of human intelligence in a relatively short period.

In many species, only males have impressive secondary sexual characteristics such as ornaments and show-off behavior, but sexual selection is also thought to be able to act on females as well in at least partially monogamous species. With complete monogamy, there is assortative mating for sexually selected traits. This means that less attractive individuals will find other less attractive individuals to mate with. If attractive traits are good fitness indicators, this means that sexual selection increases the genetic load of the offspring of unattractive individuals. Without sexual selection, an unattractive individual might find a superior mate with few deleterious mutations, and have healthy children that are likely to survive. With sexual selection, an unattractive individual is more likely to have access only to an inferior mate who is likely to pass on many deleterious mutations to their joint offspring, who are then less likely to survive.

Sexual selection is often thought to be a likely explanation for other female-specific human traits, for example breasts and buttocks far larger in proportion to total body size than those found in related species of ape. It is often assumed that if breasts and buttocks of such large size were necessary for functions such as suckling infants, they would be found in other species. That human female breasts (typical mammalian breast tissue is small) are found sexually attractive by many men is in agreement with sexual selection acting on human females secondary sexual characteristics.

Sexual selection for intelligence and judging ability can act on indicators of success, such as highly visible displays of wealth. Growing human brains require more nutrition than brains of related species of ape. It is possible that for females to successfully judge male intelligence, they must be intelligent themselves. This could explain why despite the absence of clear differences in intelligence between males and females on average, there are clear differences between male and female propensities to display their intelligence in ostentatious forms.

Critique

The sexual selection by the disability principle/fitness display model of the evolution of human intelligence is criticized by certain researchers for issues of timing of the costs relative to reproductive age. While sexually selected ornaments such as peacock feathers and moose antlers develop either during or after puberty, timing their costs to a sexually mature age, human brains expend large amounts of nutrients building myelin and other brain mechanisms for efficient communication between the neurons early in life. These costs early in life build facilitators that reduce the cost of neuron firing later in life, and as a result the peaks of the brain's costs and the peak of the brain's performance are timed on opposite sides of puberty with the costs peaking at a sexually immature age while performance peaks at a sexually mature age. Critical researchers argue that this means that the costs that intelligence is a signal of reduce the chances of surviving to reproductive age, does not signal fitness of sexually mature individuals and, since the disability principle is about selection for disabilities in sexually immature individuals that evolutionarily increase the offspring's chance of surviving to reproductive age, would be selected against and not for by its mechanisms. These critics argue that human intelligence evolved by natural selection citing that unlike sexual selection, natural selection have produced many traits that cost the most nutrients before puberty including immune systems and accumulation and modification for increased toxicity of poisons in the body as a protective measure against predators.

Intelligence as a Disease-resistance Sign

A 2008 study argues that human cleverness is simply selected within the context of sexual selection as an honest signal of genetic resistance against parasites and pathogens. The number of people with severe cognitive impairment caused by childhood viral infections like meningitis, protists like *Toxoplasma* and *Plasmodium*, and animal parasites like intestinal worms and schistosomes is estimated to be in the hundreds of millions. Even more people live with moderate mental damages, such as inability to complete difficult tasks, that are not classified as 'diseases' by medical standards, may still be considered as inferior mates by potential sexual partners.

Thus, widespread, virulent, and archaic infections are greatly involved in natural selection for cognitive abilities. People infected with parasites may have brain damage and obvious maladaptive behavior in addition to visible signs of disease. Smarter people can more skillfully learn to distinguish safe non-polluted water and food from unsafe kinds and learn to distinguish mosquito infested areas from safe areas. Smarter people can more skillfully find and develop safe food sources and living environments. Given this situation, preference for smarter child-bearing/rearing partners increases the chance that their descendants will inherit the best resistance alleles, not only for immune system resistance to disease, but also smarter brains for learning skills in avoiding disease and selecting nutritious food. When people search for mates based on their success, wealth, reputation, disease-free body appearance, or psychological traits such as benevolence or confidence; the effect is to select for superior intelligence that results in superior disease resistance.

Ecological Dominance-social Competition Model

A predominant model describing the evolution of human intelligence is ecological dominance-social competition (EDSC), explained by Mark V. Flinn, David C. Geary and Carol V. Ward based mainly on work by Richard D. Alexander. According to the model, human intelligence was able to evolve to significant levels because of the combination of increasing domination over habitat and increasing importance of social interactions. As a result, the primary selective pressure for increasing human intelligence shifted from learning to master the natural world to competition for dominance among members or groups of its own species.

As advancement, survival and reproduction within an increasing complex social structure favored ever more advanced social skills, communication of concepts through increasingly complex language patterns ensued. Since competition had shifted bit by bit from controlling "nature" to influencing other humans, it became of relevance to outmaneuver other members of the group seeking leadership or acceptance, by means of more advanced social skills. A more social and communicative person would be more easily selected.

Intelligence Dependent on Brain Size

Human intelligence is developed to an extreme level that is not necessarily adaptive in an evolutionary sense. Firstly, larger-headed babies are more difficult to give birth to and large brains are costly in terms of nutrient and oxygen requirements. Thus the direct adaptive benefit of human intelligence is questionable at least in modern societies, while it is difficult to study in prehistoric societies. Since 2005, scientists have been evaluating genomic data on gene variants thought to influence head size, and have found no evidence that those genes are under strong selective pressure in current human populations.The trait of head size has become generally fixed in modern human beings.

While decreased brain size has strong correlation with lower intelligence in humans, some modern humans have brain sizes as small as Homo Erectus but normal intelligence (based on IQ tests) for modern humans. Increased brain size in humans may allow for greater capacity for specialized expertise.

Group Selection

Group selection theory contends that organism characteristics that provide benefits to a group (clan, tribe, or larger population) can evolve despite individual disadvantages such as those cited above. The group benefits of intelligence (including language, the ability to communicate between individuals, the ability to teach others, and other cooperative aspects) have apparent utility in increasing the survival potential of a group.

Nutritional Status

Higher cognitive functioning develops better in an environment with adequate nutrition, and diets deficient in iron, zinc, protein, iodine, B vitamins, omega 3 fatty acids, magnesium and other nutrients can result in lower intelligence either in the mother during pregnancy or in the child during development. While these inputs did not have an effect on the evolution of intelligence they do govern its expression. A higher intelligence could be a signal that an individual comes from and lives in a physical and social environment where nutrition levels are high, whereas a lower intelligence could imply a child, its mother, or both, come from a physical and social environment where nutritional levels are low. Previc emphasizes the contribution of nutritional factors, especially meat and shellfish consumption, to elevations of dopaminergicactivity in the brain, which may have been responsible for the evolution of human intelligence since dopamine is crucial to working memory, cognitive shifting, abstract, distant concepts, and other hallmarks of advanced intelligence.

Behavioral Modernity

About 50,000 years ago, a major change had occurred in human behavior, described as behavioral modernity, the great leap forward, or the mind's big bang. Suddenly, the number of artifacts humans could make had exploded, suggesting an increase in behavioral plasticity.

One of the major issues in palaeoanthropology and archaeology is when our hominin ancestors became like us. Humans living today have developed the capacity for 'modern behavior'. Modern behavior can be recognized by creative and innovative culture, language, art, religious beliefs, and complex technologies (d'Errico & Stringer 2011). One of the evolved capabilities underlying modern behavior is the ability to communicate habitually and effortlessly in symbols. The pervasiveness of symbolism in present-day human culture is the reason why archaeologists frequently search for artifacts that reflect 'symbolically mediated behavior' (Henshilwood & Marean 2003). Modern behavior also consists of other components such as advanced problem solving and long range planning abilities (Wynn & Coolidge 2011). Archaeological artifacts that were produced by thinking ahead of future actions, anticipating problems and preparing responses provide evidence for modern planning abilities (Wadley 2010).

Traces of 'Modern Behavior' in the Archaeological Record

There is lively debate and divergent points of view on the appropriate markers for modern behavior in the archaeological record (Nowell 2010). Many discussions do not use theoretically developed standards such as symbolism or planning capabilities to identify modern behavior, but rely on the archaeological record itself as guide. Modern behavior has, for example, been inferred from certain traits in the archaeological record. These traits include standardization in artifact types, blade technology, worked bone and other organic materials, personal ornaments and art or images, structured living spaces, ritual, economic intensification, enlarged geographic ranges and expanded exchange networks (McBrearty & Brooks 2000). The relatively sudden appearance of such traits as a group or package in the archaeological record of the European Upper Palaeolithic has been interpreted as evidence for the onset of modern behavior (Klein 2008). This trait list approach to identify the origins of modern behavior has been criticized because the archaeological record of only one region, the Upper Palaeolithic of Europe, is used as a standard to infer modern behavior for all other time periods and areas (Deacon 1979, Henshilwood & Marean 2003, Shea 2011).

When the standard of symbolism is applied, it can be shown that artifacts of a clearly symbolic nature appear only after 100,000 years ago (Henshilwood *et al.* 2002, Henshilwood *et al.* 2004, d'Errico *et al.* 2009, Texier *et al.* 2010). These artifacts include beads as well as ochre and ostrich eggshell with geometrically engraved patterns. Obvious symbolic artifacts do not occur consistently in the archaeological record between 100,000 and 50,000 years ago, and disappear periodically (Hovers & Belfar-Cohen 2006). The fluctuating presence of symbolic artifacts may be related to changing climate and its effect on population sizes. Or perhaps it is only when ornaments with culturally dictated, three dimensional form, figurative art, depictions of mythical imagery, and musical instruments, associated with the Upper Palaeolithic of Eurasia appear, that ancient people truly became like us (Conard 2008, 2010). The absence of symbolic artifacts does not necessarily indicate the absence of symbolic capacities. It may be that ancient people used rituals, such as scarification and body painting, and objects and ornaments of perishable material that did not leave traces in the archaeological record to express their symbolic intentions. It should also be kept in mind that the presence or absence of various modern behavioral traits can be ascribed to climatic change, group size, and cultural exchange rates, rather than a lack of capacity for modernity (Richerson *et al.* 2009; Powell *et al.* 2009).

The evolution of modern planning capabilities can be investigated by analyzing the decision steps used to produce ancient tools. There are, for example, many different ways of making stone artifacts. Over the last 200,000 years a variety of reduction techniques in different combinations were used to make stone tools. This resulted in different techno-complexes, but all with the same degree of complexity. This may mean that humans had essentially modern cognitive capabilities for this period of time (Shea 2011). Using variability in stone tools as a marker has the advantage of including the most ubiquitous and widely studied type of material culture in the archaeological record in the search for modern behavior (Nowell & Davidson 2010). Advanced planning abilities were necessary to produce some ancient hunting weapons. For at least the past 200,000 years hunting implements were made by mounting sharp stone tools on a shaft with the aid of adhesives. The backed artefacts from the Howiesons Poort, dating to around 65,000 years ago in South Africa, are examples of such hafted stone tools.

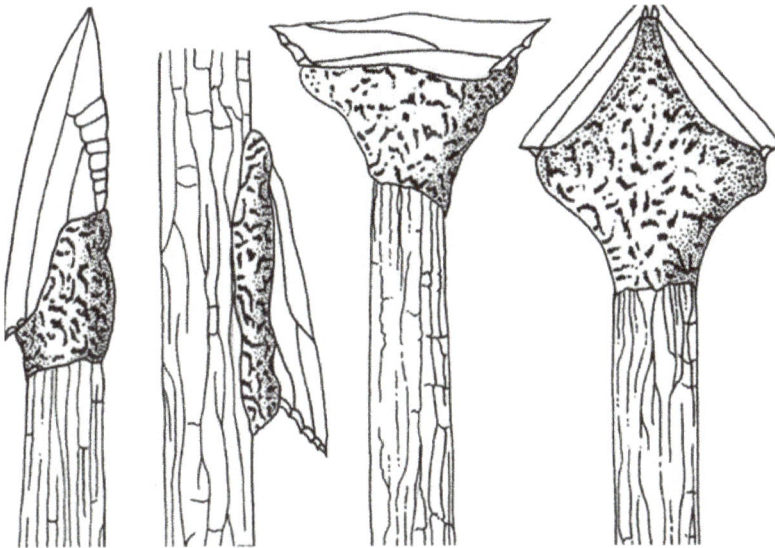

Possible hafting arrangements of Howiesons Poort backed artefacts. From left to right: diagonally hafted as tips or cutting barbs, transversal and back to back hafting

Microscopic analysis of Howiesons Poort backed tools from Sibudu Cave in KwaZulu Natal, South Africa, revealed hafting microtraces associated with microresidues of ochre and acacia gum (Lombard 2008). Lyn Wadley and her team experimentally produced adhesives using these materials, and tested how to haft stone tools to shafts. They found that several ingredients, including ochre, plant gum, and fatty substances were combined to produce the compound adhesive. In addition, several procedures and complicated use of fire, or pyrotechnology, had to be manipulated to achieve the appropriate consistency for an adhesive that would successfully join the tool to the haft when thrown. The cognitive strategy used to produce and manipulate the adhesive in the Howiesons Poort 65,000 years ago indicates multitasking and planning capabilities typical of modern people.

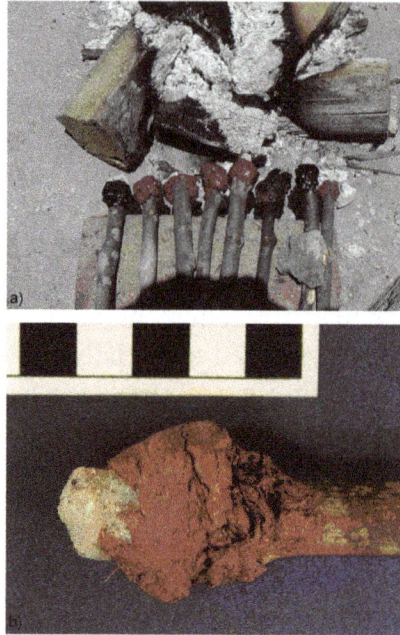

Experimental production and controlled heating of compound
adhesive, incorporating ochre, acacia gum, and fat

There is no straightforward relationship between the appearance of particular archaeological signals of modern behavior and certain kinds of fossils. Some see the development of modern behaviors as a late phenomenon, dating to 50,000 years ago, not related to the speciation of anatomically modern humans (Klein 2009). However, many archaeologists associate the development
of modern behavior with anatomically modern humans or *Homo sapiens* who emerged in Africa
around 200,000 years ago, during the Middle Stone age period. The picture is not clear, as modern
behaviors have also been linked to the Neanderthals of Europe and hominin ancestors living prior
to 200,000 years ago (Deacon & Wurz 2001, Zilhão 2007, d'Errico *et al.* 2009, d'Errico & Stringer
2011).

When considering when and how modern behavior developed, it is essential to take into account
how the brain could have evolved to support modern capabilities. Modern cognitive capacities
depend, in part, on an evolved, specialized neural architecture. Unfortunately changes in brain
size and shape that can be inferred from hominin fossils provide no clear evidence for the origins
of modern behavior. Evolutionary biology and neuroscience studies suggest that hominin symbolic communicative capabilities co-evolved with the brain, resulting in some parts of the brain
becoming proportionally larger (Deacon 1997). Humans have a larger prefrontal cortex than other
primates and this probably enabled some of the neural connections necessary for generating abstract symbolic concepts and planning tasks. The evolution of human complex functional neural
organization may have been a long-term process, involving at least a million years (Deacon 2010).
There is also the view that modern neural organization is the result of a relatively sudden genetic
mutation that took place in populations from Africa only 50,000 years ago (Klein & Edgar 2002).
Research into the origins of modern behavior must also be integrated with theories from cognitive
science in which the relationship between brain architecture and cognitive function is investigated
(Davidson 2010). An example is the interpretation of archaeological artifacts from the perspec-
tive of the executive function and working memory model (Wynn & Coolidge 2011). This model

explains how complex cognitive tasks such as planning and learning rely on the mind's ability to temporarily focus on, store, and manage information. According to this particular theory, a set of interlinked capabilities evolved, probably after 100,000 years ago, to allow enhanced working memory.

Blombos and the Origins of Modern Behavior

At the beginning of this millennium, the unexpected discovery of personal ornaments and art older than 70,000 years ago — shell beads and engraved ochre — at Blombos Cave on the southern Cape coast of South Africa, initiated a paradigm shift in thinking about the origins of modern behavior. Prior to this discovery, most scientists considered the birthplace of humans with symbolic capabilities as Upper Palaeolithic Europe (~40,000 years ago), because the earliest art and personal ornaments were found there. The finds from Blombos encouraged scientists to reconsider earlier suggestions that modern behaviors were also associated with the Middle Stone Age of Africa.

Blombos Cave yielded a large collection of tick shell (*Nassarius kraussianus*) beads — forty-nine intentionally perforated shell beads were found in layers dating to 77,000 years ago (Figure below). The shell beads were excavated in clusters of 2 to 17, and each cluster was similar in size, shade, use-wear pattern, and perforation size. The shell walls were pierced through the opening with a sharp bone point and then strung and worn. Some of the beads have traces of ochre inside and on the worn facets that could have been caused by a piercing instrument covered in ochre, or rubbing against an ochre covered skin (d'Errico *et al.* 2005). Perforated marine shells similar to those from Blombos have also been found in caves from North Africa and the Middle East. Shell beads are of the earliest evidence for personal ornaments, dating to between 100,000 - 70,000 years ago (d'Errico *et al.* 2009).

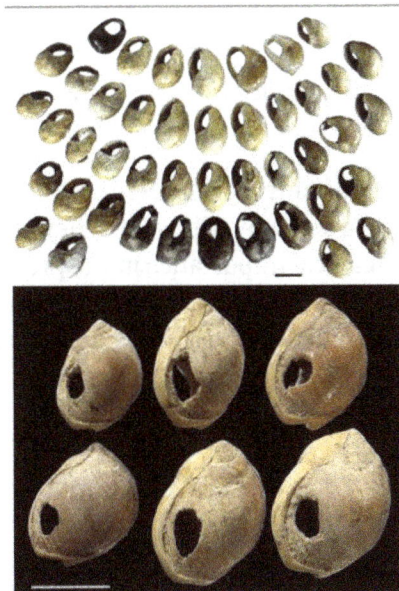

Nassarius kraussianus beads from the 77,000 year old layers at Blombos Cave, South Africa

The earliest evidence for abstract designs also comes from Blombos Cave. The most impressive find is a dark red rectangular slab of ochre (75.8mm x 34.8 mm x 24.7 mm) with a complex crosshatched engraved design. The pattern was created using a stone tool by initially engraving the longer parallel lines, whereafter the oblique lines were produced. The last step was to engrave the superficial single line that crosses the oblique lines perpendicularly. Other engraved ochres have been found in earlier layers of Blombos Cave, emphasizing this site's remarkable potential to understand symbolic behaviors of the past (Henshilwood *et al.* 2009).

The extraordinary discoveries from Blombos Cave and the other examples discussed here show that the most fruitful ways of identifying modern behavior in the archaeological record have been through artifacts that demonstrate symbolism and complex planning abilities. The challenge for future research is to expand archaeological criteria for modern behavior that are fully integrated with neuro-evolutionary theory and cognitive science.

A slab of ochre with a cross-hatched designed, evidence for abstract cognition

Human Genetics

Human genetics is the study of the inheritance of characteristics by children from parents. Inheritance in humans does not differ in any fundamental way from that in other organisms.

The study of human heredity occupies a central position in genetics. Much of this interest stems from a basic desire to know who humans are and why they are as they are. At a more practical level, an understanding of human heredity is of critical importance in the prediction, diagnosis, and treatmentof diseases that have a genetic component. The quest to determine the genetic basis of humanhealth has given rise to the field of medical genetics. In general, medicine has given focus and purpose to human genetics, so the terms *medical genetics* and *human genetics* are often considered synonymous.

DNA

DNA is perhaps the most famous biological molecule; it is present in all forms of life on earth.

Virtually every cell in your body contains DNA or the genetic code that makes you *you*. DNA carries the instructions for the development, growth, reproduction, and functioning of all life.

Differences in the genetic code are the reason why one person has blue eyes rather than brown, why some people are susceptible to certain diseases, why birds only have two wings, and why giraffes have long necks.

Amazingly, if all of the DNA in the human body was unraveled, it would reach to the sun and back more than 300 times.

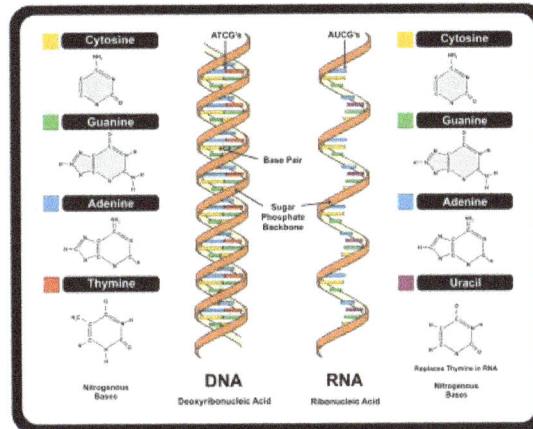

The structure of DNA and RNA. DNA is a double helix, while RNA is a single helix. Both have sets of nucleotides that contain genetic information

Deoxyribonucleic acid or DNA is a molecule that contains the instructions an organism needs to develop, live and reproduce. These instructions are found inside every cell, and are passed down from parents to their children.

Structure

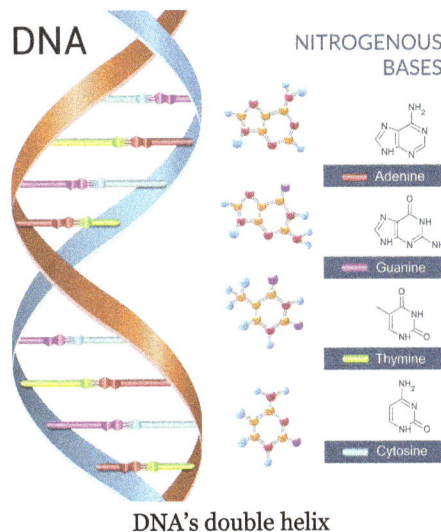

DNA's double helix

DNA is a two-stranded molecule that appears twisted, giving it a unique shape referred to as the double helix.

Each of the two strands is a long sequence of nucleotides or individual units made of:

- a phosphate molecule
- a sugar molecule called deoxyribose, containing five carbons
- a nitrogen-containing region

There are four types of nitrogen-containing regions called bases:

- adenine (A)

- cytosine (C)

- guanine (G)

- thymine (T)

The order of these four bases forms the genetic code, which is our instructions for life.

The bases of the two strands of DNA are stuck together to create a ladder-like shape. Within the ladder, A always sticks to T, and G always sticks to C to create the "rungs." The length of the ladder is formed by the sugar and phosphate groups.

Similar to the way the order of letters in the alphabet can be used to form a word, the order of nitrogen bases in a DNA sequence forms genes, which in the language of the cell, tells cells how to make proteins. Another type of nucleic acid, ribonucleic acid, or RNA, translates genetic information from DNA into proteins.

Nucleotides are attached together to form two long strands that spiral to create a structure called a double helix. If you think of the double helix structure as a ladder, the phosphate and sugar molecules would be the sides, while the bases would be the rungs. The bases on one strand pair with the bases on another strand: adenine pairs with thymine, and guanine pairs with cytosine.

Properties

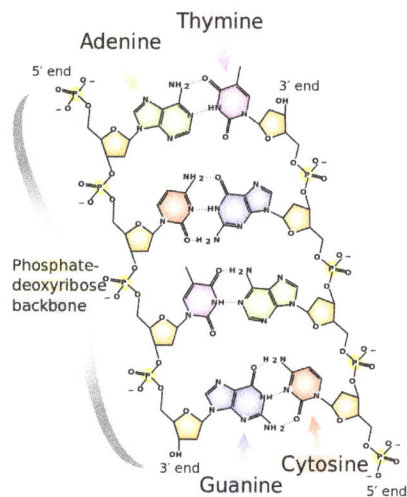

Chemical structure of DNA; hydrogen bondsshown as dotted lines

DNA is a long polymer made from repeating units called nucleotides. The structure of DNA is dynamic along its length, being capable of coiling into tight loops, and other shapes. In all species it is composed of two helical chains, bound to each other by hydrogen bonds. Both chains are coiled round the same axis, and have the same pitch of 34 ångströms(3.4 nanometres). The pair of chains has a radius of 10 ångströms (1.0 nanometre). According to another study, when measured in a different solution, the DNA chain measured 22 to 26 ångströms wide (2.2 to 2.6 nanometres), and one nucleotide unit measured 3.3 Å (0.33 nm) long. Although each individual nucleotide repeating unit is very small, DNA polymers can be very large molecules containing millions to hundreds of millions of nucleotides. For instance, the DNA in the largest human chromosome, chromosome number 1, consists of approximately 220 million base pairs and would be 85 mm long if straightened.

In living organisms, DNA does not usually exist as a single molecule, but instead as a pair of molecules that are held tightly together. These two long strands entwine like vines, in the shape of a double helix. The nucleotide contains both a segment of the backbone of the molecule (which holds the chain together) and a nucleobase (which interacts with the other DNA strand in the helix). A nucleobase linked to a sugar is called a nucleoside and a base linked to a sugar and one or more phosphate groups is called a nucleotide. A polymer comprising multiple linked nucleotides (as in DNA) is called a polynucleotide.

The backbone of the DNA strand is made from alternating phosphate and sugar residues. The sugar in DNA is 2-deoxyribose, which is a pentose (five-carbon) sugar. The sugars are joined together by phosphate groups that form phosphodiester bonds between the third and fifth carbon atoms of adjacent sugar rings, which are known as the 3′ and 5′ carbons, the prime symbol being used to distinguish these carbon atoms from those of the base to which the deoxyribose forms a glycosidic bond. When imagining DNA, each phosphoryl is normally considered to "belong" to the nucleotide whose 5′ carbon forms a bond therewith. Any DNA strand therefore normally has one end at which there is a phosphoryl attached to the 5′ carbon of a ribose (the 5′ phosphoryl) and another end at which there is a free hydroxyl attached to the 3′ carbon of a ribose (the 3′ hydroxyl). The orientation of the 3′ and 5′ carbons along the sugar-phosphate backbone confers directionality (sometimes called polarity) to each DNA strand. In a double helix, the direction of the nucleotides in one strand is opposite to their direction in the other strand: the strands are *antiparallel*. The asymmetric ends of DNA strands are said to have a directionality of *five prime* (5′) and *three prime* (3′), with the 5′ end having a terminal phosphate group and the 3′ end a terminal hydroxyl group. One major difference between DNA and RNA is the sugar, with the 2-deoxyribose in DNA being replaced by the alternative pentose sugar ribose in RNA.

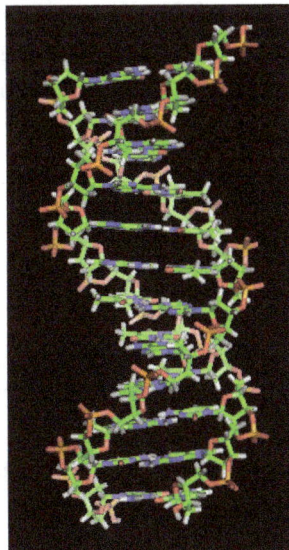

A section of DNA. The bases lie horizontally between the two spiraling strands(animated version)

The DNA double helix is stabilized primarily by two forces: hydrogen bonds between nucleotides and base-stacking interactions among aromatic nucleobases. In the aqueous environment of the cell, the conjugated π bonds of nucleotide bases align perpendicular to the axis of the DNA molecule, minimizing their interaction with the solvation shell. The four bases found in DNA are adenine (A), cytosine (C), guanine (G) and thymine (T). These four bases are attached to the sugar-phosphate to form the complete nucleotide, as shown for adenosine monophosphate. Adenine pairs with thymine and guanine pairs with cytosine. It was represented by A-T base pairs and G-C base pairs.

Nucleobase Classification

The nucleobases are classified into two types: the purines, A and G, being fused five- and six-membered heterocyclic compounds, and the pyrimidines, the six-membered rings C and T. A fifth pyrimidine nucleobase, uracil (U), usually takes the place of thymine in RNA and differs from thymine by lacking a methyl group on its ring. In addition to RNA and DNA, many artificial nucleic acid analogues have been created to study the properties of nucleic acids, or for use in biotechnology.

Non Canonical Bases

Uracil is not usually found in DNA, occurring only as a breakdown product of cytosine. However, in several bacteriophages, *Bacillus subtilis* bacteriophages PBS1 and PBS2 and *Yersinia* bacteriophage piR1-37, thymine has been replaced by uracil. Another phage - Staphylococcal phage S6 - has been identified with a genome where thymine has been replaced by uracil.

5-hydroxymethyldeoxyuridine,(hm5dU) is also known to replace thymidine in several genomes including the *Bacillus* phages SPO1, φe, SP8, H1, 2C and SP82. Another modified uracil - 5-dihydroxypentauracil – has also been described.

Base J (beta-d-glucopyranosyloxymethyluracil), a modified form of uracil, is also found in several organisms: the flagellates *Diplonema* and *Euglena*, and all the kinetoplastid genera. Biosynthesis of J occurs in two steps: in the first step, a specific thymidine in DNA is converted into hydroxymethyldeoxyuridine; in the second, HOMedU is glycosylated to form J. Proteins that bind specifically to this base have been identified. These proteins appear to be distant relatives of the Tet1 oncogene that is involved in the pathogenesis of acute myeloid leukemia. J appears to act as a termination signal for RNA polymerase II.

In 1976 a bacteriophage - S-2L - which infects species of the genus *Synechocystis* was found to have all the adenosine bases within its genome replaced by 2,6-diaminopurine. In 2016 deoxyarchaeosine was found to be present in the genomes of several bacteria and the *Escherichia* phage 9g.

Modified bases also occur in DNA. The first of these recognised was 5-methylcytosine which was found in the genome of *Mycobacterium tuberculosis* in 1925. The complete replacement of cytosine by 5-glycosylhydroxymethylcytosine in T even phages (T2, T4 and T6) was observed in 1953 In the genomes of Xanthomonas oryzae bacteriophage Xp12 and halovirus FH the full complement of cystosine has been replaced by 5-methylcytosine. 6N-methyadenine was discovered to be present in DNA in 1955. N6-carbamoyl-methyladenine was described in 1975. 7-methylguanine was described in 1976. N4-methylcytosine in DNA was described in 1983. In 1985 5-hydroxycytosine was found in the genomes of the Rhizobium phages RL38JI and N17. α-putrescinylthymine occurs in both the genomes of the *Delftia* phage ΦW-14 and the *Bacillus* phage SP10. α-glutamylthymidine is found in the Bacillus phage SP01 and 5-dihydroxypentyluracil is found in the Bacillus phage SP15.

The reason for the presence of these non canonical bases in DNA is not known. It seems likely that at least part of the reason for their presence in bacterial viruses (phages) is to avoid the restriction enzymes present in bacteria. This enzyme system acts at least in part as a molecular immune system protecting bacteria from infection by viruses.

This does not appear to be the entire story. Four modifications to the cytosine residues in human DNA have been reported. These modifications are the addition of methyl (CH_3)-, hydroxymethyl (CH_2OH)-, formyl (CHO)- and carboxyl (COOH)- groups. These modifications are thought to have regulatory functions.

Uracil is found in the centromeric regions of at least two human chromosomes (6 and 11).

Listing of Non Canonical bases Found in DNA

Seventeen non canonical bases are known to occur in DNA. Most of these are modifications of the canonical bases plus uracil.

- Modified Adenosine
  - *N6-carbamoyl-methyladenine*
  - *N6-methyadenine*
- Modified Guanine
  - *7-Methylguanine*
- Modified Cytosine
  - *N4-Methylcytosine*
  - *5-Carboxylcytosine*
  - *5-Formylcytosine*
  - *5-Glycosylhydroxymethylcytosine*
  - *5-Hydroxycytosine*
  - *5-Methylcytosine*
- Modified Thymidine
  - *α-Glutamythymidine*
  - *α-Putrescinylthymine*
- Uracil and modifications
  - *Base J*

     o   *Uracil*

     o   *5-Dihydroxypentaurauracil*

     o   *5-Hydroxymethyldeoxyuracil*

- Others

     o   *Deoxyarchaeosine*

     o   *2,6-Diaminopurine*

Grooves

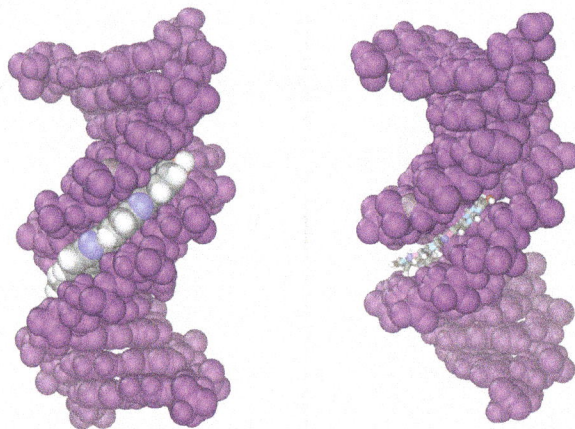

DNA major and minor grooves. The latter is a binding site for the Hoechst stain dye 33258

Twin helical strands form the DNA backbone. Another double helix may be found tracing the spaces, or grooves, between the strands. These voids are adjacent to the base pairs and may provide a binding site. As the strands are not symmetrically located with respect to each other, the grooves are unequally sized. One groove, the major groove, is 22 Å wide and the other, the minor groove, is 12 Å wide. The width of the major groove means that the edges of the bases are more accessible in the major groove than in the minor groove. As a result, proteins such as transcription factors that can bind to specific sequences in double-stranded DNA usually make contact with the sides of the bases exposed in the major groove. This situation varies in unusual conformations of DNA within the cell, but the major and minor grooves are always named to reflect the differences in size that would be seen if the DNA is twisted back into the ordinary B form.

Base Pairing

In a DNA double helix, each type of nucleobase on one strand bonds with just one type of nucleobase on the other strand. This is called complementary base pairing. Here, purines form hydrogen bondsto pyrimidines, with adenine bonding only to thymine in two hydrogen bonds, and cytosine bonding only to guanine in three hydrogen bonds. This arrangement of two nucleotides binding together across the double helix is called a Watson-Crick base pair. Another type of base pairing is Hoogsteen base pairing where two hydrogen bonds form between guanine and cytosine. As hydrogen bonds are not covalent, they can be broken and rejoined relatively easily. The two strands of DNA in a double helix can thus be pulled apart like a zipper, either by a mechanical force or high temperature. As a result of this base pair complementarity, all the information in the double-stranded sequence of a DNA helix is duplicated on each strand, which is vital in DNA replication. This reversible and specific interaction between complementary base pairs is critical for all the functions of DNA in living organisms.

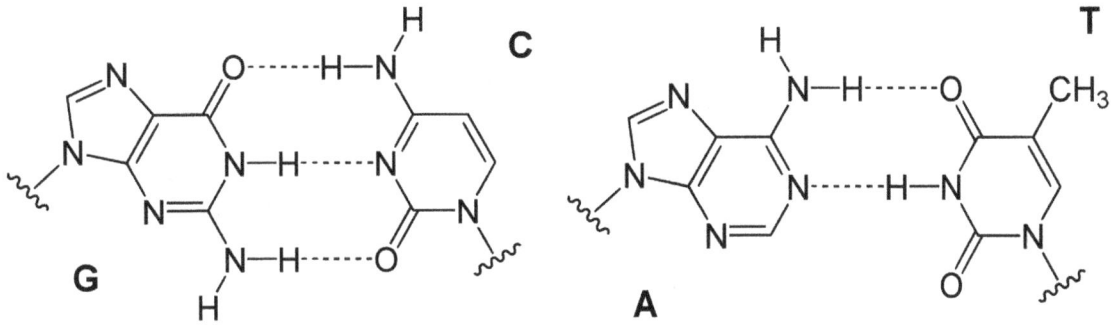

Top, a GC base pair with three hydrogen bonds. Bottom, an **AT** base pair with two hydrogen bonds. Noncovalent hydrogen bonds between the pairs are shown as dashed lines

The two types of base pairs form different numbers of hydrogen bonds, AT forming two hydrogen bonds, and GC forming three hydrogen bonds. DNA with high GC-content is more stable than DNA with low GC-content.

As noted above, most DNA molecules are actually two polymer strands, bound together in a helical fashion by noncovalent bonds; this double-stranded (dsDNA) structure is maintained largely by the intrastrand base stacking interactions, which are strongest for G, C stacks. The two strands can come apart – a process known as melting – to form two single-stranded DNA (ssDNA) molecules. Melting occurs at high temperature, low salt and high pH (low pH also melts DNA, but since DNA is unstable due to acid depurination, low pH is rarely used).

The stability of the dsDNA form depends not only on the GC-content (% G, C basepairs) but also on sequence (since stacking is sequence specific) and also length (longer molecules are more stable). The stability can be measured in various ways; a common way is the "melting temperature", which is the temperature at which 50% of the ds molecules are converted to ss molecules; melting temperature is dependent on ionic strength and the concentration of DNA. As a result, it is both the percentage of GC base pairs and the overall length of a DNA double helix that determines the strength of the association between the two strands of DNA. Long DNA helices with a high GC-content have stronger-interacting strands, while short helices with high AT content have weaker-interacting strands. In biology, parts of the DNA double helix that need to separate easily, such as the TATAAT Pribnow box in some promoters, tend to have a high AT content, making the strands easier to pull apart.

In the laboratory, the strength of this interaction can be measured by finding the temperature necessary to break the hydrogen bonds, their melting temperature (also called T_m value). When all the base pairs in a DNA double helix melt, the strands separate and exist in solution as two entirely independent molecules. These single-stranded DNA molecules have no single common shape, but some conformations are more stable than others.

Sense and Antisense

A DNA sequence is called "sense" if its sequence is the same as that of a messenger RNA copy that is translated into protein. The sequence on the opposite strand is called the "antisense" sequence. Both sense and antisense sequences can exist on different parts of the same strand of DNA (i.e. both strands can contain both sense and antisense sequences). In both prokaryotes and eukaryotes, antisense RNA sequences are produced, but the functions of these RNAs are not entirely clear. One proposal is that antisense RNAs are involved in regulating gene expression through RNA-RNA base pairing.

A few DNA sequences in prokaryotes and eukaryotes, and more in plasmids and viruses, blur the distinction between sense and antisense strands by having overlapping genes. In these cases, some DNA sequences do double duty, encoding one protein when read along one strand, and a second protein when read in the opposite direction along the other strand. In bacteria, this overlap may be involved in the regulation of gene transcription, while in viruses, overlapping genes increase the amount of information that can be encoded within the small viral genome.

Supercoiling

DNA can be twisted like a rope in a process called DNA supercoiling. With DNA in its "relaxed" state, a strand usually circles the axis of the double helix once every 10.4 base pairs, but if the DNA is twisted the strands become more tightly or more loosely wound. If the DNA is twisted in the direction of the helix, this is positive supercoiling, and the bases are held more tightly together. If they are twisted in the opposite direction, this is negative supercoiling, and the bases come apart more easily. In nature, most DNA has slight negative supercoiling that is introduced by enzymes called topoisomerases. These enzymes are also needed to relieve the twisting stresses introduced into DNA strands during processes such as transcription and DNA replication.

Alternative DNA Structures

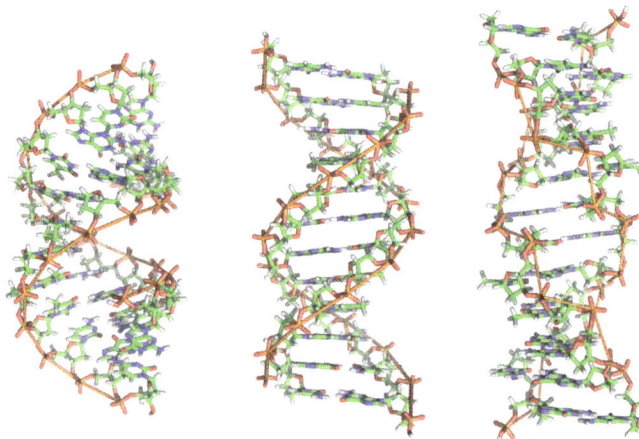

From left to right, the structures of A, B and Z DNA

DNA exists in many possible conformations that include A-DNA, B-DNA, and Z-DNA forms, although, only B-DNA and Z-DNA have been directly observed in functional organisms. The conformation that DNA adopts depends on the hydration level, DNA sequence, the amount and direction of supercoiling, chemical modifications of the bases, the type and concentration of metal ions, and the presence of polyamines in solution.

The first published reports of A-DNA X-ray diffraction patterns—and also B-DNA—used analyses based on Patterson transforms that provided only a limited amount of structural information for oriented fibers of DNA. An alternative analysis was then proposed by Wilkins *et al.*, in 1953, for the *in vivo* B-DNA X-ray diffraction-scattering patterns of highly hydrated DNA fibers in terms of squares of Bessel functions. In the same journal, James Watson and Francis Crick presented their molecular modeling analysis of the DNA X-ray diffraction patterns to suggest that the structure was a double-helix.

Although the *B-DNA form* is most common under the conditions found in cells, it is not a well-defined conformation but a family of related DNA conformations that occur at the high hydration levels present in living cells. Their corresponding X-ray diffraction and scattering patterns are characteristic of molecular paracrystals with a significant degree of disorder.

Compared to B-DNA, the A-DNA form is a wider right-handed spiral, with a shallow, wide minor groove and a narrower, deeper major groove. The A form occurs under non-physiological conditions in partly dehydrated samples of DNA, while in the cell it may be produced in hybrid pairings of DNA and RNA strands, and in enzyme-DNA complexes. Segments of DNA where the bases have been chemically modified by methylation may undergo a larger change in conformation and adopt the Z form. Here, the strands turn about the helical axis in a left-handed spiral, the opposite of the more common B form. These unusual structures can be recognized by specific Z-DNA binding proteins and may be involved in the regulation of transcription.

Alternative DNA Chemistry

For many years exobiologists have proposed the existence of a shadow biosphere, a postulated microbial biosphere of Earth that uses radically different biochemical and molecular processes than currently known life. One of the proposals was the existence of lifeforms that use arsenic instead of phosphorus in DNA. A report in 2010 of the possibility in the bacterium GFAJ-1, was announced, though the research was disputed, and evidence suggests the bacterium actively prevents the incorporation of arsenic into the DNA backbone and other biomolecules.

Quadruplex Structures

At the ends of the linear chromosomes are specialized regions of DNA called telomeres. The main function of these regions is to allow the cell to replicate chromosome ends using the enzyme telomerase, as the enzymes that normally replicate DNA cannot copy the extreme 3′ ends of chromosomes. These specialized chromosome caps also help protect the DNA ends, and stop the DNA repair systems in the cell from treating them as damage to be corrected. In human cells, telomeres are usually lengths of single-stranded DNA containing several thousand repeats of a simple TTAGGG sequence.

These guanine-rich sequences may stabilize chromosome ends by forming structures of stacked sets of four-base units, rather than the usual base pairs found in other DNA molecules. Here, four guanine bases form a flat plate and these flat four-base units then stack on top of each other, to form a stable G-quadruplex structure. These structures are stabilized by hydrogen bonding between the edges of the bases and chelation of a metal ion in the centre of each four-base unit. Other structures can also be formed, with the central set of four bases coming from either a single strand folded around the bases, or several different parallel strands, each contributing one base to the central structure.

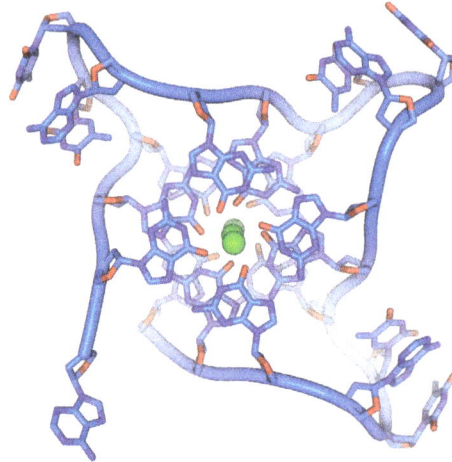

DNA quadruplex formed by telomere repeats. The looped conformation of the DNA backbone is very different from the typical DNA helix. The green spheres in the center represent potassium ions

In addition to these stacked structures, telomeres also form large loop structures called telomere loops, or T-loops. Here, the single-stranded DNA curls around in a long circle stabilized by telomere-binding proteins. At the very end of the T-loop, the single-stranded telomere DNA is held onto a region of double-stranded DNA by the telomere strand disrupting the double-helical DNA and base pairing to one of the two strands. This triple-stranded structure is called a displacement loop or D-loop.

Branched DNA

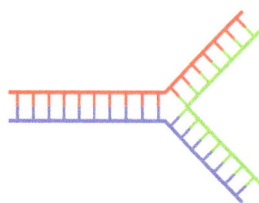

Single branch   Multiple branches
Branched DNA can form networks containing multiple branches

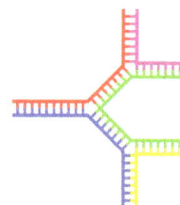

In DNA, fraying occurs when non-complementary regions exist at the end of an otherwise complementary double-strand of DNA. However, branched DNA can occur if a third strand of DNA is introduced and contains adjoining regions able to hybridize with the frayed regions of the pre-existing double-strand. Although the simplest example of branched DNA involves only three strands of DNA, complexes involving additional strands and multiple branches are also possible. Branched DNA can be used in nanotechnology to construct geometric shapes.

Chemical Modifications and Altered DNA Packaging

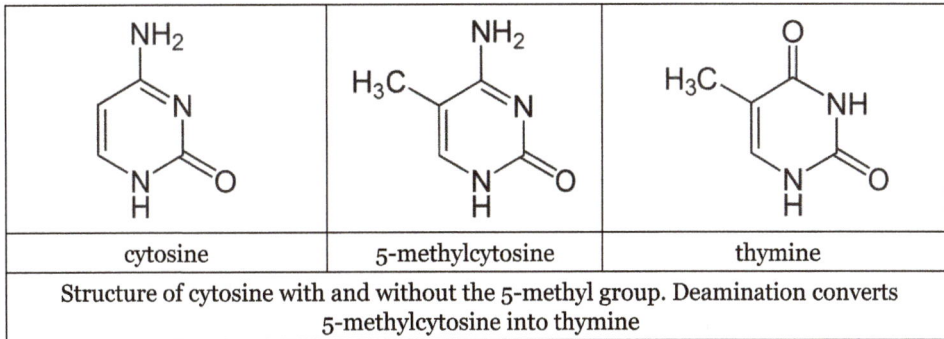

cytosine	5-methylcytosine	thymine
Structure of cytosine with and without the 5-methyl group. Deamination converts 5-methylcytosine into thymine		

Base Modifications and DNA Packaging

The expression of genes is influenced by how the DNA is packaged in chromosomes, in a structure called chromatin. Base modifications can be involved in packaging, with regions that have low or no gene expression usually containing high levels of methylation of cytosine bases. DNA packaging and its influence on gene expression can also occur by covalent modifications of the histone protein core around which DNA is wrapped in the chromatin structure or else by remodeling carried out by chromatin remodeling complexes. There is, further, crosstalk between DNA methylation and histone modification, so they can coordinately affect chromatin and gene expression.

For one example, cytosine methylation produces 5-methylcytosine, which is important for X-inactivation of chromosomes. The average level of methylation varies between organisms – the worm *Caenorhabditis elegans* lacks cytosine methylation, while vertebrates have higher levels, with up to 1% of their DNA containing 5-methylcytosine. Despite the importance of 5-methylcytosine, it can deaminate to leave a thymine base, so methylated cytosines are particularly prone to mutations. Other base modifications include adenine methylation in bacteria, the presence of 5-hydroxymethylcytosine in the brain, and the glycosylation of uracil to produce the "J-base" in kinetoplastids.

Damage

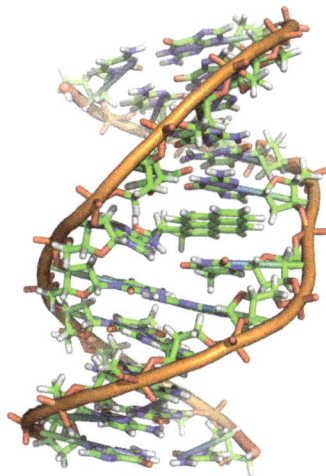

A covalent adduct between a metabolically activated form of benzo[*a*]pyrene, the major mutagen in tobacco smoke, and DNA

DNA can be damaged by many sorts of mutagens, which change the DNA sequence. Mutagens include oxidizing agents, alkylating agents and also high-energy electromagnetic radiationsuch as ultraviolet light and X-rays. The type of DNA damage produced depends on the type of mutagen. For example, UV light can damage DNA by producing thymine dimers, which are cross-links between pyrimidine bases. On the other hand, oxidants such as free radicals or hydrogen peroxide produce multiple forms of damage, including base modifications, particularly of guanosine, and double-strand breaks. A typical human cell contains about 150,000 bases that have suffered oxidative damage. Of these oxidative lesions, the most dangerous are double-strand breaks, as these are difficult to repair and can produce point mutations, insertions, deletions from the DNA sequence, and chromosomal translocations. These mutations can cause cancer. Because of inherent limits in the DNA repair mechanisms, if humans lived long enough, they would all eventually develop cancer. DNA damages that are naturally occurring, due to normal cellular processes that produce reactive oxygen species, the hydrolytic activities of cellular water, etc., also occur frequently. Although most of these damages are repaired, in any cell some DNA damage may remain despite the action of repair processes. These remaining DNA damages accumulate with age in mammalian postmitotic tissues. This accumulation appears to be an important underlying cause of aging.

Many mutagens fit into the space between two adjacent base pairs, this is called *intercalation*. Most intercalators are aromatic and planar molecules; examples include ethidium bromide, acridines, daunomycin, and doxorubicin. For an intercalator to fit between base pairs, the bases must separate, distorting the DNA strands by unwinding of the double helix. This inhibits both transcription and DNA replication, causing toxicity and mutations. As a result, DNA intercalators may be carcinogens, and in the case of thalidomide, a teratogen. Others such as benzo[a]pyrene diol epoxide and aflatoxin form DNA adducts that induce errors in replication. Nevertheless, due to their ability to inhibit DNA transcription and replication, other similar toxins are also used in chemotherapy to inhibit rapidly growing cancer cells.

Biological Functions

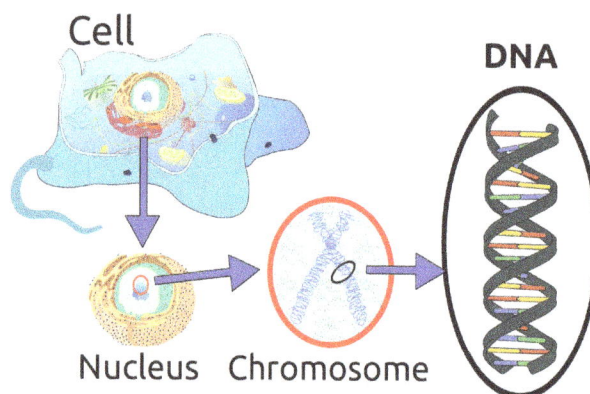

Location of eukaryote nuclear DNA within the chromosomes

DNA usually occurs as linear chromosomes in eukaryotes, and circular chromosomes in prokaryotes. The set of chromosomes in a cell makes up its genome; the human genome has approximately 3 billion base pairs of DNA arranged into 46 chromosomes. The information carried by DNA is held in the sequence of pieces of DNA called genes. Transmission of genetic information in genes is achieved via complementary base pairing. For example, in transcription, when a cell uses the information in a gene, the DNA sequence is copied into a complementary RNA sequence through the attraction between the DNA and the correct RNA nucleotides. Usually, this RNA copy is then used to make a matching protein sequence in a process called translation, which depends on the same interaction between RNA nucleotides. In alternative fashion, a cell may simply copy its genetic information in a process called DNA replication. The details of these functions are covered in other articles; here the focus is on the interactions between DNA and other molecules that mediate the function of the genome.

Genes and Genomes

Genomic DNA is tightly and orderly packed in the process called DNA condensation, to fit the small available volumes of the cell. In eukaryotes, DNA is located in the cell nucleus, with small amounts in mitochondria and chloroplasts. In prokaryotes, the DNA is held within an irregularly shaped body in the cytoplasm called the nucleoid. The genetic information in a genome is held within genes, and the complete set of this information in an organism is called its genotype. A gene is a unit of heredity and is a region of DNA that influences a particular characteristic in an organism. Genes contain an open reading frame that can be transcribed, and regulatory sequences such as promoters and enhancers, which control transcription of the open reading frame.

In many species, only a small fraction of the total sequence of the genome encodes protein. For example, only about 1.5% of the human genome consists of protein-coding exons, with over 50% of human DNA consisting of non-coding repetitive sequences. The reasons for the presence of so much noncoding DNA in eukaryotic genomes and the extraordinary differences in genome size, or *C-value*, among species, represent a long-standing puzzle known as the "C-value enigma". However, some DNA sequences that do not code protein may still encode functional non-coding RNA molecules, which are involved in the regulation of gene expression.

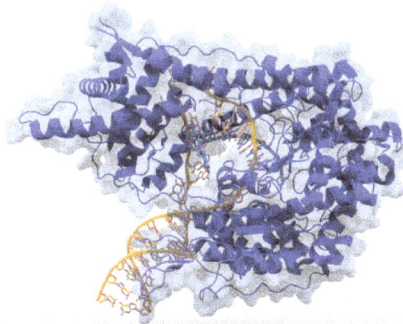

T7 RNA polymerase (blue) producing an mRNA (green) from a DNA template (orange)

Some noncoding DNA sequences play structural roles in chromosomes. Telomeres and centromeres typically contain few genes but are important for the function and stability of chromosomes. An abundant form of noncoding DNA in humans are pseudogenes, which are copies of genes that have been disabled by mutation. These sequences are usually just molecular fossils, although they can occasionally serve as raw genetic material for the creation of new genes through the process of gene duplication and divergence.

Transcription and Translation

A gene is a sequence of DNA that contains genetic information and can influence the phenotype of an organism. Within a gene, the sequence of bases along a DNA strand defines a messenger RNA-sequence, which then defines one or more protein sequences. The relationship between the nucleotide sequences of genes and the amino-acid sequences of proteins is determined by the rules of translation, known collectively as the genetic code. The genetic code consists of three-letter 'words' called *codons* formed from a sequence of three nucleotides (e.g. ACT, CAG, TTT).

In transcription, the codons of a gene are copied into messenger RNA by RNA polymerase. This RNA copy is then decoded by a ribosome that reads the RNA sequence by base-pairing the messenger RNA to transfer RNA, which carries amino acids. Since there are 4 bases in 3-letter combinations, there are 64 possible codons (4^3 combinations). These encode the twenty standard amino acids, giving most amino acids more than one possible codon. There are also three 'stop' or 'nonsense' codons signifying the end of the coding region; these are the TAA, TGA, and TAG codons.

Replication

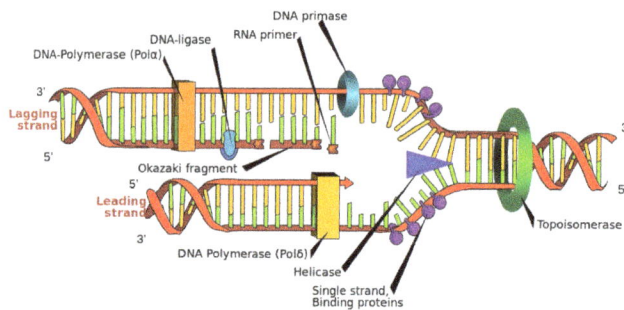

DNA replication. The double helix is unwound by a helicase and topoisomerase. Next, one DNA polymerase produces the leading strand copy. Another DNA polymerase binds to the lagging strand. This enzyme makes discontinuous segments (called Okazaki fragments) before DNA ligase joins them together

Cell division is essential for an organism to grow, but, when a cell divides, it must replicate the DNA in its genome so that the two daughter cells have the same genetic information as their parent. The double-stranded structure of DNA provides a simple mechanism for DNA replication. Here, the two strands are separated and then each strand's complementary DNA sequence is recreated by an enzyme called DNA polymerase. This enzyme makes the complementary strand by finding the correct base through complementary base pairing and bonding it onto the original strand. As DNA polymerases can only extend a DNA strand in a 5′ to 3′ direction, different mechanisms are used to copy the antiparallel strands of the double helix. In this way, the base on the old strand dictates which base appears on the new strand, and the cell ends up with a perfect copy of its DNA.

Extracellular Nucleic Acids

Naked extracellular DNA (eDNA), most of it released by cell death, is nearly ubiquitous in the environment. Its concentration in soil may be as high as 2 µg/L, and its concentration in natural aquatic environments may be as high at 88 µg/L. Various possible functions have been proposed for eDNA: it may be involved in horizontal gene transfer; it may provide nutrients; and it may act as a buffer to recruit or titrate ions or antibiotics. Extracellular DNA acts as a functional extracellular matrix component in the biofilms of several bacterial species. It may act as a recognition factor to regulate the attachment and dispersal of specific cell types in the biofilm; it may contribute to biofilm formation; and it may contribute to the biofilm's physical strength and resistance to biological stress.

Cell-free fetal DNA is found in the blood of the mother, and can be sequenced to determine a great deal of information about the developing fetus.

Interactions with Proteins

All the functions of DNA depend on interactions with proteins. These protein interactions can be non-specific, or the protein can bind specifically to a single DNA sequence. Enzymes can also bind to DNA and of these, the polymerases that copy the DNA base sequence in transcription and DNA replication are particularly important.

DNA-binding Proteins

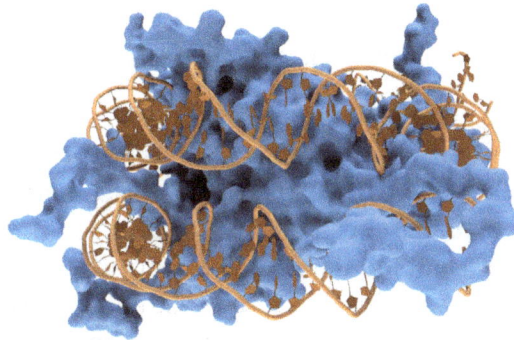

Interaction of DNA (in orange) with histones (in blue). These proteins' basic amino acids bind to the acidic phosphate groups on DNA

Structural proteins that bind DNA are well-understood examples of non-specific DNA-protein interactions. Within chromosomes, DNA is held in complexes with structural proteins. These proteins organize the DNA into a compact structure called chromatin. In eukaryotes, this structure involves DNA binding to a complex of small basic proteins called histones, while in prokaryotes multiple types of proteins are involved. The histones form a disk-shaped complex called a nucleosome, which contains two complete turns of double-stranded DNA wrapped around its surface. These non-specific interactions are formed through basic residues in the histones, making ionic bonds to the acidic sugar-phosphate backbone of the DNA, and are thus largely independent of the base sequence. Chemical modifications of these basic amino acid residues include methylation, phosphorylation, and acetylation. These chemical changes alter the strength of the interaction between the DNA and the histones, making the DNA more or less accessible to transcription factors and changing the rate of transcription. Other non-specific DNA-binding proteins in chromatin include the high-mobility group proteins, which bind to bent or distorted DNA. These proteins are important in bending arrays of nucleosomes and arranging them into the larger structures that make up chromosomes.

A distinct group of DNA-binding proteins is the DNA-binding proteins that specifically bind single-stranded DNA. In humans, replication protein A is the best-understood member of this family and is used in processes where the double helix is separated, including DNA replication, recombination, and DNA repair. These binding proteins seem to stabilize single-stranded DNA and protect it from forming stem-loops or being degraded by nucleases.

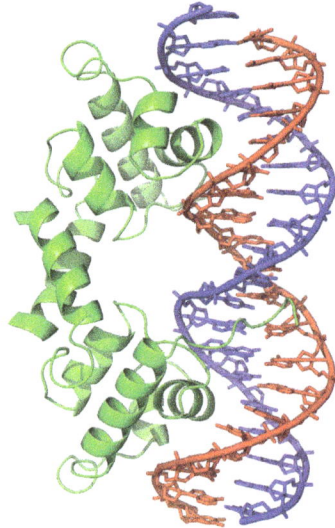

The lambda repressor helix-turn-helix transcription factor bound to its DNA target

In contrast, other proteins have evolved to bind to particular DNA sequences. The most intensively studied of these are the various transcription factors, which are proteins that regulate transcription. Each transcription factor binds to one particular set of DNA sequences and activates or inhibits the transcription of genes that have these sequences close to their promoters. The transcription factors do this in two ways. Firstly, they can bind the RNA polymerase responsible for transcription, either directly or through other mediator proteins; this locates the polymerase at the promoter and allows it to begin transcription. Alternatively, transcription factors can bind enzymes that modify the histones at the promoter. This changes the accessibility of the DNA template to the polymerase.

As these DNA targets can occur throughout an organism's genome, changes in the activity of one type of transcription factor can affect thousands of genes. Consequently, these proteins are often the targets of the signal transduction processes that control responses to environmental changes or cellular differentiation and development. The specificity of these transcription factors' interactions with DNA come from the proteins making multiple contacts to the edges of the DNA bases, allowing them to "read" the DNA sequence. Most of these base-interactions are made in the major groove, where the bases are most accessible.

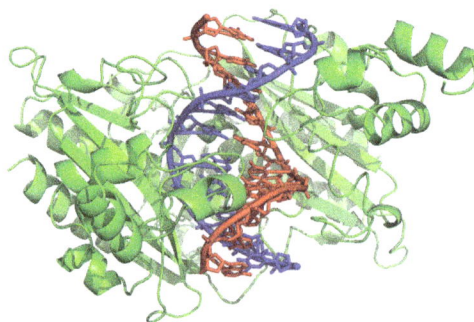

The restriction enzyme EcoRV(green) in a complex with its substrate DNA

DNA-modifying Enzymes

Nucleases and Ligases

Nucleases are enzymes that cut DNA strands by catalyzing the hydrolysis of the phosphodiester bonds. Nucleases that hydrolyse nucleotides from the ends of DNA strands are called exonucleases, while endonucleases cut within strands. The most frequently used nucleases in molecular biology are the restriction endonucleases, which cut DNA at specific sequences. For instance, the EcoRV enzyme shown to the left recognizes the 6-base sequence 5′-GATATC-3′ and makes a cut at the horizontal line. In nature, these enzymes protect bacteria against phage infection by digesting the phage DNA when it enters the bacterial cell, acting as part of the restriction modification system. In technology, these sequence-specific nucleases are used in molecular cloning and DNA fingerprinting.

Enzymes called DNA ligases can rejoin cut or broken DNA strands. Ligases are particularly important in lagging strand DNA replication, as they join together the short segments of DNA produced at the replication fork into a complete copy of the DNA template. They are also used in DNA repair and genetic recombination.

Topoisomerases and Helicases

Topoisomerases are enzymes with both nuclease and ligase activity. These proteins change the amount of supercoiling in DNA. Some of these enzymes work by cutting the DNA helix and allowing one section to rotate, thereby reducing its level of supercoiling; the enzyme then seals the DNA break. Other types of these enzymes are capable of cutting one DNA helix and then passing a second strand of DNA through this break, before rejoining the helix. Topoisomerases are required for many processes involving DNA, such as DNA replication and transcription.

Helicases are proteins that are a type of molecular motor. They use the chemical energy in nucleoside triphosphates, predominantly adenosine triphosphate (ATP), to break hydrogen bonds between bases and unwind the DNA double helix into single strands. These enzymes are essential for most processes where enzymes need to access the DNA bases.

Polymerases

Polymerases are enzymes that synthesize polynucleotide chains from nucleoside triphosphates. The sequence of their products is created based on existing polynucleotide chains—which are called *templates*. These enzymes function by repeatedly adding a nucleotide to the 3′ hydroxyl group at the end of the growing polynucleotide chain. As a consequence, all polymerases work in a 5′ to 3′ direction. In the active site of these enzymes, the incoming nucleoside triphosphate base-pairs to the template: this allows polymerases to accurately synthesize the complementary strand of their template. Polymerases are classified according to the type of template that they use.

In DNA replication, DNA-dependent DNA polymerases make copies of DNA polynucleotide chains. To preserve biological information, it is essential that the sequence of bases in each copy are precisely complementary to the sequence of bases in the template strand. Many DNA polymerases have a proofreading activity. Here, the polymerase recognizes the occasional mistakes in the synthesis reaction by the lack of base pairing between the mismatched nucleotides. If a mismatch is detected, a 3′ to 5′ exonuclease activity is activated and the incorrect base removed. In most organisms, DNA polymerases function in a large complex called the replisome that contains multiple accessory subunits, such as the DNA clamp or helicases.

RNA-dependent DNA polymerases are a specialized class of polymerases that copy the sequence of an RNA strand into DNA. They include reverse transcriptase, which is a viral enzyme involved in the infection of cells by retroviruses, and telomerase, which is required for the replication of telomeres. For example, HIV reverse transcriptase is an enzyme for AIDS virus replication. Telomerase is an unusual polymerase because it contains its own RNA template as part of its structure. It synthesizes telomeres at the ends of chromosomes. Telomeres prevent fusion of the ends of neighboring chromosomes and protect chromosome ends from damage.

Transcription is carried out by a DNA-dependent RNA polymerase that copies the sequence of a DNA strand into RNA. To begin transcribing a gene, the RNA polymerase binds to a sequence of DNA called a promoter and separates the DNA strands. It then copies the gene sequence into a messenger RNA transcript until it reaches a region of DNA called the terminator, where it halts and detaches from the DNA. As with human DNA-dependent DNA polymerases, RNA polymerase II, the enzyme that transcribes most of the genes in the human genome, operates as part of a large protein complex with multiple regulatory and accessory subunits.

Genetic Recombination

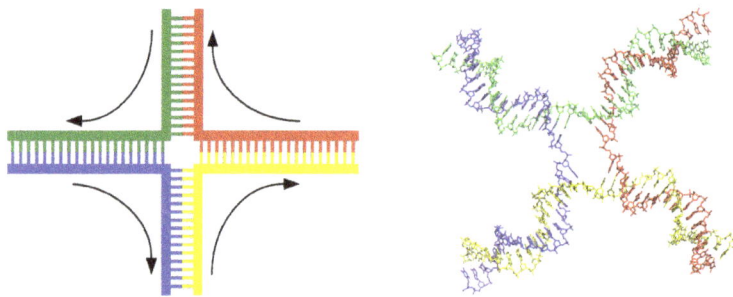

Structure of the Holliday junctionintermediate in genetic recombination.
The four separate DNA strands are coloured red, blue, green and yellow

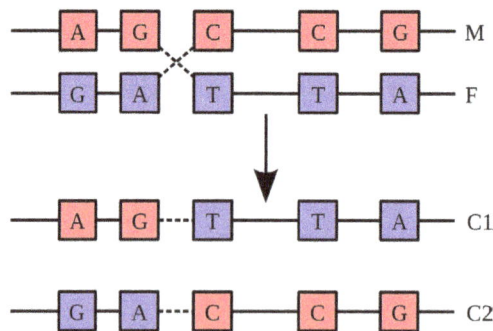

Recombination involves the breaking and rejoining of two chromosomes (M and F) to produce two rearranged chromosomes (C1 and C2)

A DNA helix usually does not interact with other segments of DNA, and in human cells, the different chromosomes even occupy separate areas in the nucleus called "chromosome territories". This physical separation of different chromosomes is important for the ability of DNA to function as a stable repository for information, as one of the few times chromosomes interact is in chromosomal crossover which occurs during sexual reproduction, when genetic recombination occurs. Chromosomal crossover is when two DNA helices break, swap a section and then rejoin.

Recombination allows chromosomes to exchange genetic information and produces new combinations of genes, which increases the efficiency of natural selection and can be important in the rapid evolution of new proteins. Genetic recombination can also be involved in DNA repair, particularly in the cell's response to double-strand breaks.

The most common form of chromosomal crossover is homologous recombination, where the two chromosomes involved share very similar sequences. Non-homologous recombination can be damaging to cells, as it can produce chromosomal translocations and genetic abnormalities. The recombination reaction is catalyzed by enzymes known as recombinases, such as RAD51. The first step in recombination is a double-stranded break caused by either an endonuclease or damage to the DNA. A series of steps catalyzed in part by the recombinase then leads to joining of the two helices by at least one Holliday junction, in which a segment of a single strand in each helix is annealed to the complementary strand in the other helix. The Holliday junction is a tetrahedral junction structure that can be moved along the pair of chromosomes, swapping one strand for another. The recombination reaction is then halted by cleavage of the junction and re-ligation of the released DNA. Only strands of like polarity exchange DNA during recombination. There are two types of cleavage: east-west cleavage and north-south cleavage. The north-south cleavage nicks both strands of DNA, while the east-west cleavage has one strand of DNA intact. The formation of a Holliday junction during recombination makes it possible for genetic diversity, genes to exchange on chromosomes, and expression of wild-type viral genomes.

Evolution

DNA contains the genetic information that allows all modern living things to function, grow and reproduce. However, it is unclear how long in the 4-billion-year history of life DNA has performed this function, as it has been proposed that the earliest forms of life may have used RNA as their genetic material. RNA may have acted as the central part of early cell metabolism as it can both transmit genetic information and carry out catalysis as part of ribozymes. This ancient RNA world where nucleic acid would have been used for both catalysis and genetics may have influenced the evolution of the current genetic code based on four nucleotide bases. This would occur, since the number of different bases in such an organism is a trade-off between a small number of bases increasing replication accuracy and a large number of bases increasing the catalytic efficiency of ribozymes. However, there is no direct evidence of ancient genetic systems, as recovery of DNA from most fossils is impossible because DNA survives in the environment for less than one million years, and slowly degrades into short fragments in solution. Claims for older DNA have been made, most notably a report of the isolation of a viable bacterium from a salt crystal 250 million years old, but these claims are controversial.

Building blocks of DNA (adenine, guanine, and related organic molecules) may have been formed extraterrestrially in outer space. Complex DNA and RNA organic compounds of life, including uracil, cytosine, and thymine, have also been formed in the laboratory under conditions mimicking those found in outer space, using starting chemicals, such as pyrimidine, found in meteorites. Pyrimidine, like polycyclic aromatic hydrocarbons (PAHs), the most carbon-rich chemical found in the universe, may have been formed in red giants or in interstellar cosmic dust and gas clouds.

RNA

RNA, ribonucleic acid is the complex compound of high molecular weight that functions in cellular protein synthesis and replaces DNA (deoxyribonucleic acid) as a carrier of genetic codes in some viruses. RNA consists of ribose nucleotides (nitrogenous bases appended to a ribose sugar) attached by phosphodiester bonds, forming strands of varying lengths. The nitrogenous bases in RNA are adenine, guanine, cytosine, and uracil, which replaces thymine in DNA.

The ribose sugar of RNA is a cyclical structure consisting of five carbons and one oxygen. The presence of a chemically reactive hydroxyl (–OH) group attached to the second carbon group in the ribose sugar molecule makes RNA prone to hydrolysis. This chemical lability of RNA, compared with DNA, which does not have a reactive –OH group in the same position on the sugar moiety (deoxyribose), is thought to be one reason why DNA evolved to be the preferred carrier of genetic information in most organisms. The structure of the RNA molecule was described by R.W. Holley in 1965.

RNA Structure

RNA typically is a single-stranded biopolymer. However, the presence of self-complementary sequences in the RNA strand leads to intrachain base-pairing and folding of the ribonucleotide chain into complex structural forms consisting of bulges and helices. The three-dimensional structure of RNA is critical to its stability and function, allowing the ribose sugar and the nitrogenous bases to be modified in numerous different ways by cellular enzymes that attach chemical groups (e.g., methyl groups) to the chain. Such modifications enable the formation of chemical bonds between distant regions in the RNA strand, leading to complex contortions in the RNA chain, which further stabilizes the RNA structure. Molecules with weak structural modifications and stabilization may be readily destroyed. As an example, in an initiator transfer RNA (tRNA) molecule that lacks a methyl group($tRNA_i^{Met}$), modification at position 58 of the tRNA chain renders the molecule unstable and hence nonfunctional; the nonfunctional chain is destroyed by cellular tRNA quality control mechanisms.

RNAs can also form complexes with molecules known as ribonucleoproteins (RNPs). The RNA portion of at least one cellular RNP has been shown to act as a biological catalyst, a function previously ascribed only to proteins.

Types of RNA

Messenger RNA (mRNA) is the RNA that carries information from DNA to the ribosome, the sites of protein synthesis (translation) in the cell. The coding sequence of the mRNA determines the amino acid sequence in the protein that is produced. However, many RNAs do not code for protein (about 97% of the transcriptional output is non-protein-coding in eukaryotes).

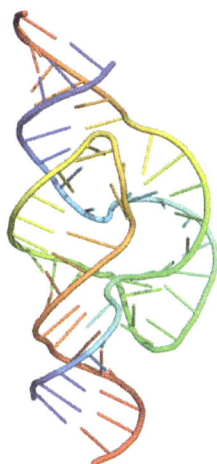

Structure of a hammerhead ribozyme, a ribozyme that cuts RNA

These so-called non-coding RNAs ("ncRNA") can be encoded by their own genes (RNA genes), but can also derive from mRNA introns. The most prominent examples of non-coding RNAs are transfer RNA (tRNA) and ribosomal RNA (rRNA), both of which are involved in the process of translation. There are also non-coding RNAs involved in gene regulation, RNA processing and other roles. Certain RNAs are able to catalyse chemical reactions such as cutting and ligating other RNA molecules, and the catalysis of peptide bond formation in the ribosome; these are known as ribozymes.

In Length

According to the length of RNA chain, RNA includes small RNA and long RNA. Usually, small RNAs are shorter than 200 nt in length, and long RNAs are greater than 200 nt long. Long RNAs, also called large RNAs, mainly include long non-coding RNA (lncRNA) and mRNA. Small RNAs mainly include 5.8S ribosomal RNA (rRNA), 5S rRNA, transfer RNA (tRNA), microRNA (miR-NA), small interfering RNA (siRNA), small nucleolar RNA (snoRNAs), Piwi-interacting RNA (piR-NA), tRNA-derived small RNA (tsRNA) and small rDNA-derived RNA (srRNA).

In Translation

Messenger RNA (mRNA) carries information about a protein sequence to the ribosomes, the protein synthesis factories in the cell. It is coded so that every three nucleotides (a codon) corresponds to one amino acid. In eukaryotic cells, once precursor mRNA (pre-mRNA) has been transcribed from DNA, it is processed to mature mRNA. This removes its introns—non-coding sections of the pre-mRNA. The mRNA is then exported from the nucleus to the cytoplasm, where it is bound to ribosomes and translated into its corresponding protein form with the help of tRNA. In prokaryotic cells, which do not have nucleus and cytoplasm compartments, mRNA can bind to ribosomes while it is being transcribed from DNA. After a certain amount of time, the message degrades into its component nucleotides with the assistance of ribonucleases.

Transfer RNA (tRNA) is a small RNA chain of about 80 nucleotides that transfers a specific amino acid to a growing polypeptide chain at the ribosomal site of protein synthesis during translation. It has sites for amino acid attachment and an anticodon region for codon recognition that binds to a specific sequence on the messenger RNA chain through hydrogen bonding.

Ribosomal RNA (rRNA) is the catalytic component of the ribosomes. Eukaryotic ribosomes contain four different rRNA molecules: 18S, 5.8S, 28S and 5S rRNA. Three of the rRNA molecules are synthesized in the nucleolus, and one is synthesized elsewhere. In the cytoplasm, ribosomal RNA and protein combine to form a nucleoprotein called a ribosome. The ribosome binds mRNA and carries out protein synthesis. Several ribosomes may be attached to a single mRNA at any time. Nearly all the RNA found in a typical eukaryotic cell is rRNA.

Transfer-messenger RNA (tmRNA) is found in many bacteria and plastids. It tags proteins encoded by mRNAs that lack stop codons for degradation and prevents the ribosome from stalling.

Regulatory RNAs

Several types of RNA can downregulate gene expression by being complementary to a part of an mRNA or a gene's DNA. MicroRNAs (miRNA; 21–22 nt) are found in eukaryotes and act through RNA interference (RNAi), where an effector complex of miRNA and enzymes can cleave complementary mRNA, block the mRNA from being translated, or accelerate its degradation.

While small interfering RNAs (siRNA; 20–25 nt) are often produced by breakdown of viral RNA, there are also endogenous sources of siRNAs. siRNAs act through RNA interference in a fashion similar to miRNAs. Some miRNAs and siRNAs can cause genes they target to be methylated, thereby decreasing or increasing transcription of those genes. Animals have Piwi-interacting RNAs (piRNA; 29–30 nt) that are active in germline cells and are thought to be a defense against transposons and play a role in gametogenesis.

Many prokaryotes have CRISPR RNAs, a regulatory system similar to RNA interference. Antisense RNAs are widespread; most downregulate a gene, but a few are activators of transcription. One way antisense RNA can act is by binding to an mRNA, forming double-stranded RNA that is enzymatically degraded. There are many long noncoding RNAs that regulate genes in eukaryotes, one such RNA is Xist, which coats one X chromosome in female mammals and inactivates it.

An mRNA may contain regulatory elements itself, such as riboswitches, in the 5' untranslated region or 3' untranslated region; these cis-regulatory elements regulate the activity of that mRNA. The untranslated regions can also contain elements that regulate other genes.

In RNA Processing

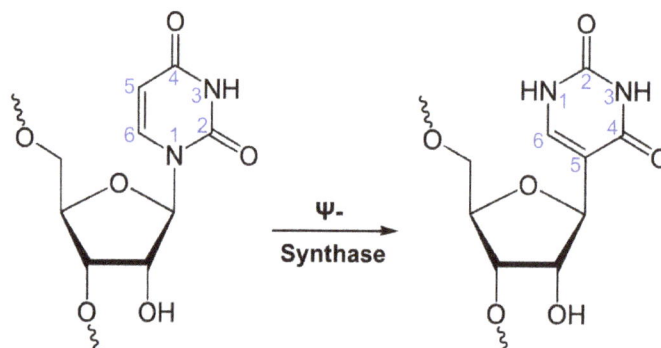

Uridine to pseudouridine is a common RNA modification

Many RNAs are involved in modifying other RNAs. Introns are spliced out of pre-mRNA by spliceosomes, which contain several small nuclear RNAs (snRNA), or the introns can be ribozymes that are spliced by themselves. RNA can also be altered by having its nucleotides modified to nucleotides other than A, C, Gand U. In eukaryotes, modifications of RNA nucleotides are in general directed by small nucleolar RNAs (snoRNA; 60–300 nt), found in the nucleolus and cajal bodies. snoRNAs associate with enzymes and guide them to a spot on an RNA by basepairing to that RNA. These enzymes then perform the nucleotide modification. rRNAs and tRNAs are extensively modified, but snRNAs and mRNAs can also be the target of base modification. RNA can also be methylated.

RNA Genomes

Like DNA, RNA can carry genetic information. RNA viruses have genomes composed of RNA that encodes a number of proteins. The viral genome is replicated by some of those proteins, while other proteins protect the genome as the virus particle moves to a new host cell. Viroids are another group of pathogens, but they consist only of RNA, do not encode any protein and are replicated by a host plant cell's polymerase.

In Reverse Transcription

Reverse transcribing viruses replicate their genomes by reverse transcribing DNA copies from their RNA; these DNA copies are then transcribed to new RNA. Retrotransposons also spread by copying DNA and RNA from one another, and telomerase contains an RNA that is used as template for building the ends of eukaryotic chromosomes.

Double-stranded RNA

Double-stranded RNA (dsRNA) is RNA with two complementary strands, similar to the DNA found in all cells, but with the replacement of thymine by uracil. dsRNA forms the genetic material of some viruses(double-stranded RNA viruses). Double-stranded RNA, such as viral RNA or siRNA, can trigger RNA interference in eukaryotes, as well as interferon response in vertebrates.

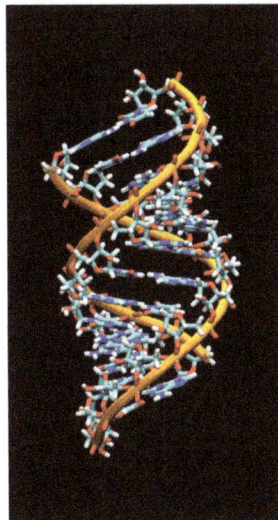

Double-stranded RNA

Circular RNA

In the late 1970s, it was shown that there is a single stranded covalently closed, i.e. circular form of RNA expressed throughout the animal and plant kingdom. circRNAs are thought to arise via a "back-splice" reaction where the spliceosome joins a downstream donor to an upstream acceptor splice site. So far the function of circRNAs is largely unknown, although for few examples a microRNA sponging activity has been demonstrated.

RNA In Disease

Important connections have been discovered between RNA and human disease. For example, as described previously, some miRNAs are capable of regulating cancer-associated genes in ways that facilitate tumour development. In addition, the dysregulation of miRNA metabolism has been linked to various neurodegenerative diseases, including Alzheimer disease. In the case of other RNA types, tRNAs can bind to specialized proteins known as caspases, which are involved in apoptosis(programmed cell death). By binding to caspase proteins, tRNAs inhibit apoptosis; the ability of cells to escape programmed death signaling is a hallmark of cancer. Noncoding RNAs known as tRNA-derived fragments (tRFs) are also suspected to play a role in cancer. The emergence of techniques such as RNA sequencing has led to the identification of novel classes of tumour-specific RNA transcripts, such as MALAT1 (metastasis associated lung adenocarcinoma transcript 1), increased levels of which have been found in various cancerous tissues and are associated with the proliferation and metastasis (spread) of tumour cells.

A class of RNAs containing repeat sequences is known to sequester RNA-binding proteins (RBPs), resulting in the formation of foci or aggregates in neural tissues. These aggregates play a role in the development of neurological diseases such as amyotrophic lateral sclerosis (ALS) and myotonic dystrophy. The loss of function, dysregulation, and mutation of various RBPs has been implicated in a host of human diseases.

The discovery of additional links between RNA and disease is expected. Increased understanding of RNA and its functions, combined with the continued development of sequencing technologies and efforts to screen RNA and RBPs as therapeutic targets, are likely to facilitate such discoveries.

Chromosome

Chromosome is the microscopic threadlike part of the cell that carries hereditary information in the form of genes. A defining feature of any chromosome is its compactness. For instance, the 46 chromosomes found in human cells have a combined length of 200 nm (1 nm = 10^{-9} metre); if the chromosomes were to be unraveled, the genetic material they contain would measure roughly 2 metres (about 6.5 feet) in length. The compactness of chromosomes plays an important role in helping to organize genetic material during cell division and enabling it to fit inside structures such as the nucleus of a cell, the average diameter of which is about 5 to 10 μm (1 μm = 0.001 mm, or 0.000039 inch), or the polygonal head of a virus particle, which may be in the range of just 20 to 30 nm in diameter.

The structure and location of chromosomes are among the chief differences between viruses, prokaryotes, and eukaryotes. The nonliving viruses have chromosomes consisting of either DNA (deoxyribonucleic acid) or RNA (ribonucleic acid); this material is very tightly packed into the viral head. Among organisms with prokaryotic cells (i.e., bacteria and blue-green algae), chromosomes consist entirely of DNA. The single chromosome of a prokaryotic cell is not enclosed within a nuclear membrane. Among eukaryotes, the chromosomes are contained in a membrane-bound cell nucleus. The chromosomes of a eukaryotic cell consist primarily of DNA attached to a protein core. They also contain RNA.

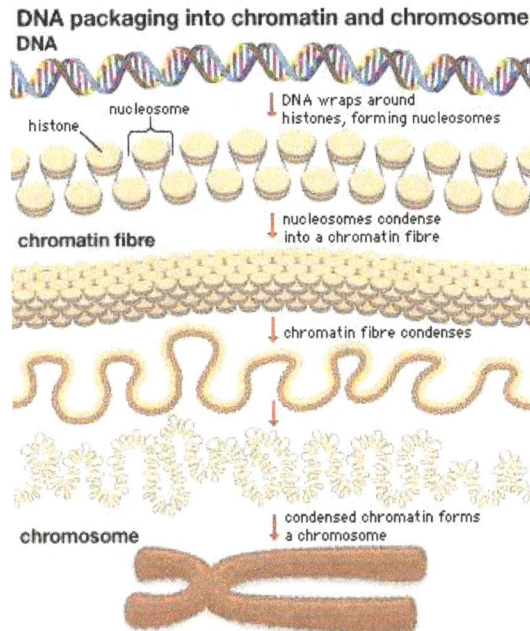

DNA packaging into chromatin and chromosome

DNA wraps around proteins called histones to form units known as nucleosomes.
These units condense into a chromatin fibre, which condenses further to form a chromosome

Every eukaryotic species has a characteristic number of chromosomes (chromosome number). In species that reproduce asexually, the chromosome number is the same in all the cells of the organism. Among sexually reproducing organisms, the number of chromosomes in the body (somatic) cells is diploid ($2n$; a pair of each chromosome), twice the haploid ($1n$) number found in the sex cells, or gametes. The haploid number is produced during meiosis. During fertilization, two gametes combine to produce a zygote, a single cell with a diploid set of chromosomes.

Somatic cells reproduce by dividing, a process called mitosis. Between cell divisions the chromosomes exist in an uncoiled state, producing a diffuse mass of genetic material known as chromatin. The uncoiling of chromosomes enables DNA synthesis to begin. During this phase, DNA duplicates itself in preparation for cell division.

Following replication, the DNA condenses into chromosomes. At this point, each chromosome actually consists of a set of duplicate chromatids that are held together by the centromere. The centromere is the point of attachment of the kinetochore, a protein structure that is connected to the spindle fibres (part of a structure that pulls the chromatids to opposite ends of the cell). During the middle stage in cell division, the centromere duplicates, and the chromatid pair separates; each chromatid becomes a separate chromosome at this point. The cell divides, and both of the daughter cells have a complete (diploid) set of chromosomes. The chromosomes uncoil in the new cells, again forming the diffuse network of chromatin.

Among many organisms that have separate sexes, there are two basic types of chromosomes: sex chromosomes and autosomes. Autosomes control the inheritance of all the characteristics except the sex-linked ones, which are controlled by the sex chromosomes. Humans have 22 pairs of autosomes and one pair of sex chromosomes. All act in the same way during cell division.

Chromosome breakage is the physical breakage of subunits of a chromosome. It is usually followed by reunion (frequently at a foreign site, resulting in a chromosome unlike the original). Breakage and reunion of homologous chromosomes during meiosis are the basis for the classical model of crossing over, which results in unexpected types of offspring of a mating.

Prokaryotes

The prokaryotes – bacteria and archaea – typically have a single circular chromosome, but many variations exist. The chromosomes of most bacteria, which some authors prefer to call genophores, can range in size from only 130,000 base pairs in the endosymbiotic bacteria *Candidatus Hodgkinia cicadicola* and *Candidatus Tremblaya princeps*, to more than 14,000,000 base pairs in the soil-dwelling bacterium *Sorangium cellulosum*. Spirochaetes of the genus *Borrelia*are a notable exception to this arrangement, with bacteria such as *Borrelia burgdorferi*, the cause of Lyme disease, containing a single *linear* chromosome.

Structure in Sequences

Prokaryotic chromosomes have less sequence-based structure than eukaryotes. Bacteria typically have a one-point (the origin of replication) from which replication starts, whereas some archaea contain multiple replication origins. The genes in prokaryotes are often organized in operons, and do not usually contain introns, unlike eukaryotes.

DNA Packaging

Prokaryotes do not possess nuclei. Instead, their DNA is organized into a structure called the nucleoid. The nucleoid is a distinct structure and occupies a defined region of the bacterial cell. This structure is, however, dynamic and is maintained and remodeled by the actions of a range of histone-like proteins, which associate with the bacterial chromosome. In archaea, the DNA in chromosomes is even more organized, with the DNA packaged within structures similar to eukaryotic nucleosomes.

Certain bacteria also contain plasmids or other extrachromosomal DNA. These are circular structures in the cytoplasm that contain cellular DNA and play a role in horizontal gene transfer. In prokaryotes and viruses, the DNA is often densely packed and organized; in the case of archaea, by homology to eukaryotic histones, and in the case of bacteria, by histone-like proteins.

Bacterial chromosomes tend to be tethered to the plasma membrane of the bacteria. In molecular biology application, this allows for its isolation from plasmid DNA by centrifugation of lysed bacteria and pelleting of the membranes (and the attached DNA).

Prokaryotic chromosomes and plasmids are, like eukaryotic DNA, generally supercoiled. The DNA must first be released into its relaxed state for access for transcription, regulation, and replication.

Eukaryotes

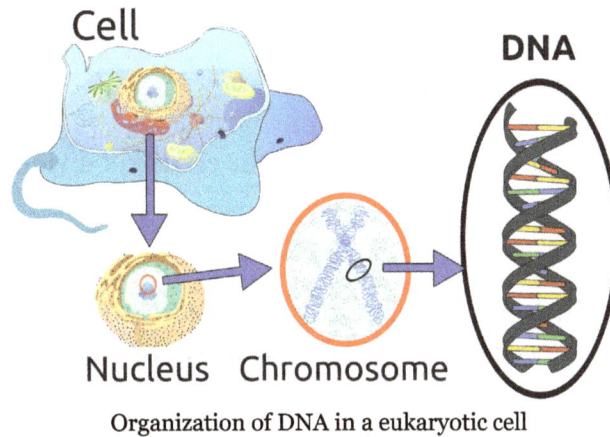

Organization of DNA in a eukaryotic cell

Chromosomes in eukaryotes are composed of chromatin fiber. Chromatin fiber is made of nucleosomes (histone octamers with part of a DNA strand attached to and wrapped around it). Chromatin fibers are packaged by proteins into a condensed structure called chromatin. Chromatin contains the vast majority of DNA and a small amount inherited maternally, can be found in the mitochondria. Chromatin is present in most cells, with a few exceptions, for example, red blood cells.

Chromatin allows the very long DNA molecules to fit into the cell nucleus. During cell division chromatin condenses further to form microscopically visible chromosomes. The structure of chromosomes varies through the cell cycle. During cellular division chromosomes are replicated, divided, and passed successfully to their daughter cells so as to ensure the genetic diversity and survival of their progeny. Chromosomes may exist as either duplicated or unduplicated. Unduplicated chromosomes are single double helixes, whereas duplicated chromosomes contain two identical copies (called chromatids or sister chromatids) joined by a centromere.

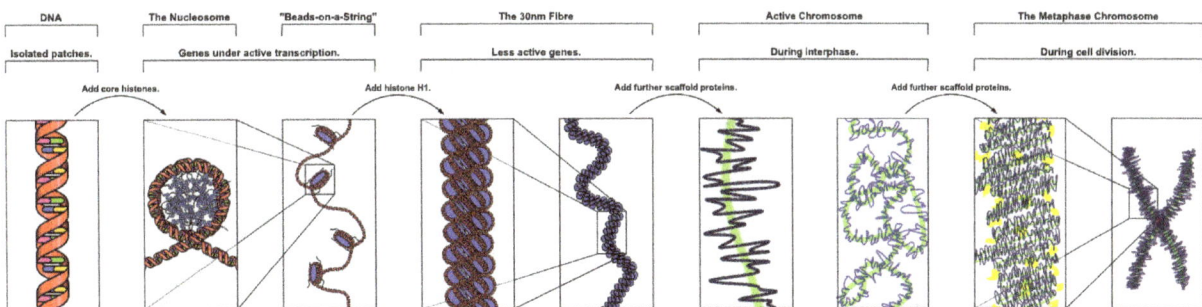

The major structures in DNA compaction: DNA, the nucleosome, the 10 nm "beads-on-a-string" fibre, the 30 nm fibre and the metaphase chromosome

Eukaryotes (cells with nuclei such as those found in plants, fungi, and animals) possess multiple large linear chromosomes contained in the cell's nucleus. Each chromosome has one centromere, with one or two arms projecting from the centromere, although, under most circumstances, these arms are not visible as such. In addition, most eukaryotes have a small circular mitochondrial genome, and some eukaryotes may have additional small circular or linear cytoplasmic-chromosomes.

In the nuclear chromosomes of eukaryotes, the uncondensed DNA exists in a semi-ordered structure, where it is wrapped around histones (structural proteins), forming a composite material called chromatin.

Interphase Chromatin

During interphase (the period of the cell cycle where the cell is not dividing), two types of chromatin can be distinguished:

- Euchromatin, which consists of DNA that is active, e.g., being expressed as protein.

- Heterochromatin, which consists of mostly inactive DNA. It seems to serve structural purposes during the chromosomal stages. Heterochromatin can be further distinguished into two types:

  o *Constitutive heterochromatin, which is never expressed. It is located around the centromere and usually contains repetitive sequences.*

  o *Facultative heterochromatin, which is sometimes expressed.*

Structure of Eukaryotic Chromosome

1. Each chromosome is made up of two chromatids(chromosomal arms) which are joined to each other at a small constricted region called the centromere. (Primary constriction). These sister chromatids are conjoined twins the result of DNA replication.

2. The centromere helps the chromatids attach to the spindle fibres during cell division, it is also concerned with the anaphase movement of the chromosomes, by which the spindle fibers pull the chromatids to the two opposite poles by their contraction during anaphase.

3. Besides the primary constriction, in certain chromosomes there is a secondary constriction as well. Because a small portion is pinched off from the chromosomal body; this portion is called a 'satellite' and the chromosome is called an SAT chromosome.

4. The two chromatids are made up of very thin chromatin fibres which are made up of 40% DNA and 60% histone proteins

5. Each chromatin fibre consists of one DNA helix coiled around eight histone molecules like a loop; such a complex is called nucleosome and resembles a bead on a string. These nucleosomes pack tighter, during condensation required to get to metaphase.

6. The primary constriction cannot take up most stains, so during cell division this region is a gap in staining.

7. Within the primary constriction there is a clear zone called Centromere.

8. The centromere with the DNA and histone proteins bound to them form a disc shaped structure called *kinetochore.*

9. The chromonemata is a word that means a chromatid in the early stage of condensation.

Metaphase Chromatin and Division

In the early stages of mitosis or meiosis (cell division), the chromatin double helix become more and more condensed. They cease to function as accessible genetic material (transcription stops) and become a compact transportable form. This compact form makes the individual chromosomes visible, and they form the classic four arm structure, a pair of sister chromatids attached to each other at the centromere. The shorter arms are called *p arms* (from the French *petit*, small) and the longer arms are called *q arms* (*q* follows *p* in the Latin alphabet; q-g "grande"; alternatively it is sometimes said q is short for *queue* meaning tail in French). This is the only natural context in which individual chromosomes are visible with an optical microscope.

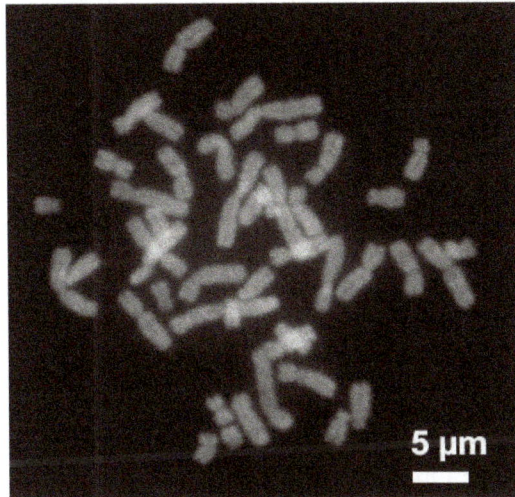

Human chromosomes during metaphase

Mitotic metaphase chromosomes are best described by a linearly organized longitudinally compressed array of consecutive chromatin loops.

During mitosis, microtubules grow from centrosomes located at opposite ends of the cell and also attach to the centromere at specialized structures called kinetochores, one of which is present on each sister chromatid. A special DNA base sequence in the region of the kinetochores provides, along with special proteins, longer-lasting attachment in this region. The microtubules then pull the chromatids apart toward the centrosomes, so that each daughter cell inherits one set of chromatids. Once the cells have divided, the chromatids are uncoiled and DNA can again be transcribed. In spite of their appearance, chromosomes are structurally highly condensed, which enables these giant DNA structures to be contained within a cell nucleus.

Human Chromosomes

Chromosomes in humans can be divided into two types: autosomes (body chromosome(s)) and allosome (sex chromosome(s)). Certain genetic traits are linked to a person's sex and are passed on through the sex chromosomes. The autosomes contain the rest of the genetic hereditary information. All act in the same way during cell division. Human cells have 23 pairs of chromosomes (22 pairs of autosomes and one pair of sex chromosomes), giving a total of 46 per cell. In addition to these, human cells have many hundreds of copies of the mitochondrial genome. Sequencing of the human genome has provided a great deal of information about each of the chromosomes.

Below is a table compiling statistics for the chromosomes, based on the Sanger Institute's human genome information in the Vertebrate Genome Annotation (VEGA) database. Number of genes is an estimate, as it is in part based on gene predictions. Total chromosome length is an estimate as well, based on the estimated size of unsequenced heterochromatin regions.

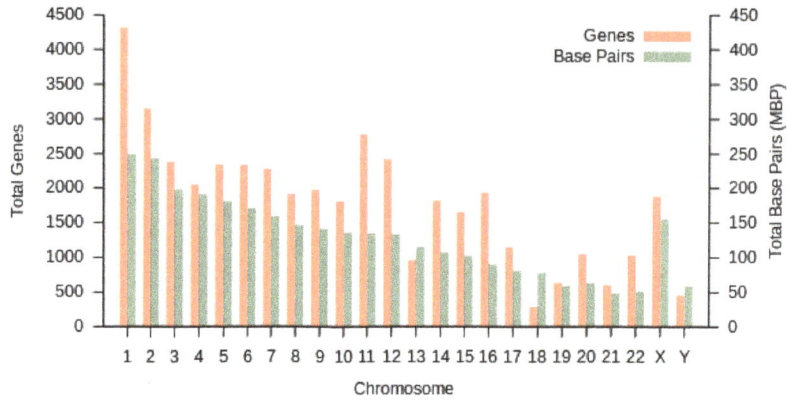

Estimated number of genes and base pairs (in mega base pairs) on each human chromosome

Chromosome	Genes	Total base pairs	% of bases	Sequenced base pairs
1	2000	247,199,719	8.0	224,999,719
2	1300	242,751,149	7.9	237,712,649
3	1000	199,446,827	6.5	194,704,827
4	1000	191,263,063	6.2	187,297,063
5	900	180,837,866	5.9	177,702,766
6	1000	170,896,993	5.5	167,273,993
7	900	158,821,424	5.2	154,952,424
8	700	146,274,826	4.7	142,612,826
9	800	140,442,298	4.6	120,312,298
10	700	135,374,737	4.4	131,624,737
11	1300	134,452,384	4.4	131,130,853
12	1100	132,289,534	4.3	130,303,534
13	300	114,127,980	3.7	95,559,980
14	800	106,360,585	3.5	88,290,585
15	600	100,338,915	3.3	81,341,915
16	800	88,822,254	2.9	78,884,754
17	1200	78,654,742	2.6	77,800,220
18	200	76,117,153	2.5	74,656,155
19	1500	63,806,651	2.1	55,785,651
20	500	62,435,965	2.0	59,505,254
21	200	46,944,323	1.5	34,171,998
22	500	49,528,953	1.6	34,893,953
X (sex chromosome)	800	154,913,754	5.0	151,058,754
Y (sex chromosome)	50	57,741,652	1.9	25,121,652
Total	21,000	3,079,843,747	100.0	2,857,698,560

Number in Various Organisms

In Eukaryotes

These tables give the total number of chromosomes (including sex chromosomes) in a cell nucleus. For example, most eukaryotes are diploid, like humans who have 22 different types of autosomes, each present as two homologous pairs, and two sex chromosomes. This gives 46 chromosomes in total. Other organisms have more than two copies of their chromosome types, such as bread wheat, which is *hexaploid* and has six copies of seven different chromosome types – 42 chromosomes in total.

Chromosome numbers in some plants	
Plant Species	
Arabidopsis thaliana (diploid)	10
Rye (diploid)	14
Einkorn wheat (diploid)	14
Maize (diploid or palaeotetraploid)	20
Durum wheat (tetraploid)	28
Bread wheat (hexaploid)	42
Cultivated tobacco (tetraploid)	48
Adder's tongue fern (polyploid)	approx. 1,200

Chromosome numbers (2n) in some animals	
Species	
Indian muntjac	7
Common fruit fly	8
Pill millipede (*Arthrosphaera fumosa*)	30
Earthworm (*Octodrilus complanatus*)	36
Tibetan fox	36
Domestic cat	38
Domestic pig	38
Laboratory mouse	40
Laboratory rat	42
Rabbit (*Oryctolagus cuniculus*)	44
Syrian hamster	44
Guppy (*poecilia reticulata*)	46
Human	46
Hares	48
Gorillas, chimpanzees	48
Domestic sheep	54
Garden snail	54
Silkworm	56
Elephants	56
Cow	60
Donkey	62
Guinea pig	64
Horse	64

Chromosome numbers (2n) in some animals	
Species	
Dog	78
Hedgehog	90
Goldfish	100-104
Kingfisher	132

Chromosome numbers in other organisms			
Species	**Large Chromosomes**	**Intermediate Chromosomes**	**Microchromosomes**
Trypanosoma brucei	11	6	~100
Domestic pigeon (*Columba livia domestics*)	18	-	59-63
Chicken	8	2 sex chromosomes	60

Normal members of a particular eukaryotic species all have the same number of nuclear chromosomes (see the table). Other eukaryotic chromosomes, i.e., mitochondrial and plasmid-like small chromosomes, are much more variable in number, and there may be thousands of copies per cell.

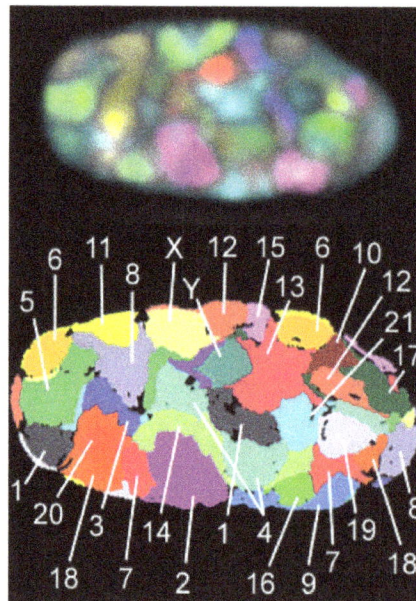

The 23 human chromosome territories during prometaphase in fibroblast cells

Asexually reproducing species have one set of chromosomes that are the same in all body cells. However, asexual species can be either haploid or diploid.

Sexually reproducing species have somatic cells (body cells), which are diploid [2n] having two sets of chromosomes (23 pairs in humans with one set of 23 chromosomes from each parent), one set from the mother and one from the father. Gametes, reproductive cells, are haploid [n]: They have one set of chromosomes. Gametes are produced by meiosis of a diploid germ line cell. During meiosis, the matching chromosomes of father and mother can exchange small parts of themselves (crossover), and thus create new chromosomes that are not inherited solely from either parent. When a male and a female gamete merge (fertilization), a new diploid organism is formed.

Some animal and plant species are polyploid [Xn]: They have more than two sets of homologous chromosomes. Plants important in agriculture such as tobacco or wheat are often polyploid, compared to their ancestral species. Wheat has a haploid number of seven chromosomes, still seen in some cultivars as well as the wild progenitors. The more-common pasta and bread wheat types are polyploid, having 28 (tetraploid) and 42 (hexaploid) chromosomes, compared to the 14 (diploid) chromosomes in the wild wheat.

In Prokaryotes

Prokaryote species generally have one copy of each major chromosome, but most cells can easily survive with multiple copies. For example, *Buchnera*, a symbiont of aphids has multiple copies of its chromosome, ranging from 10–400 copies per cell. However, in some large bacteria, such as *Epulopiscium fishelsoni* up to 100,000 copies of the chromosome can be present. Plasmids and plasmid-like small chromosomes are, as in eukaryotes, highly variable in copy number. The number of plasmids in the cell is almost entirely determined by the rate of division of the plasmid – fast division causes high copy number.

Karyotype

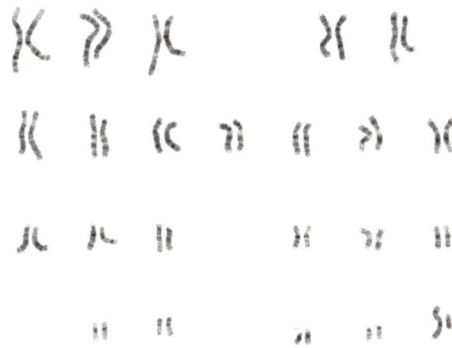

Karyogram of a human male

In general, the karyotype is the characteristic chromosome complement of a eukaryote species. The preparation and study of karyotypes is part of cytogenetics.

Although the replication and transcription of DNA is highly standardized in eukaryotes, *the same cannot be said for their karyotypes*, which are often highly variable. There may be variation between species in chromosome number and in detailed organization. In some cases, there is significant variation within species. Often there is:

1. variation between the two sexes

2. variation between the germ-line and soma (between gametes and the rest of the body)

3. variation between members of a population, due to balanced genetic polymorphism

4. geographical variation between races

5. mosaics or otherwise abnormal individuals.

Also, variation in karyotype may occur during development from the fertilized egg.

The technique of determining the karyotype is usually called *karyotyping*. Cells can be locked part-way through division (in metaphase) in vitro (in a reaction vial) with colchicine. These cells are then stained, photographed, and arranged into a *karyogram*, with the set of chromosomes arranged, autosomes in order of length, and sex chromosomes (here X/Y) at the end.

Like many sexually reproducing species, humans have special gonosomes (sex chromosomes, in contrast to autosomes). These are XX in females and XY in males.

Investigation into the human karyotype took many years to settle the most basic question: *How many chromosomes does a normal diploid human cell contain?* In 1912, Hans von Winiwarter reported 47 chromosomes in spermatogonia and 48 in oogonia, concluding an XX/XO sex determination mechanism. Painter in 1922 was not certain whether the diploid number of man is 46 or 48, at first favouring 46. He revised his opinion later from 46 to 48, and he correctly insisted on humans having an XX/XY system.

New techniques were needed to definitively solve the problem:

1. Using cells in culture

2. Arresting mitosis in metaphase by a solution of colchicine

3. Pretreating cells in a hypotonic solution 0.075 M KCl, which swells them and spreads the chromosomes

4. Squashing the preparation on the slide forcing the chromosomes into a single plane

5. Cutting up a photomicrograph and arranging the result into an indisputable karyogram.

It took until 1954 before the human diploid number was confirmed as 46. Considering the techniques of Winiwarter and Painter, their results were quite remarkable. Chimpanzees, the closest living relatives to modern humans, have 48 chromosomes as do the other great apes: in humans two chromosomes fused to form chromosome 2.

Aberrations

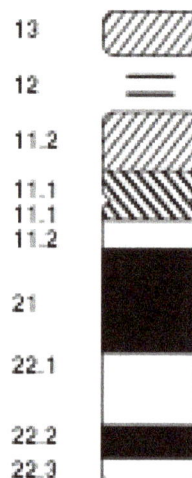

In Down syndrome, there are three copies of chromosome 21

Chromosomal aberrations are disruptions in the normal chromosomal content of a cell and are a major cause of genetic conditions in humans, such as Down syndrome, although most aberrations have little to no effect. Some chromosome abnormalities do not cause disease in carriers, such as translocations, or chromosomal inversions, although they may lead to a higher chance of bearing a child with a chromosome disorder. Abnormal numbers of chromosomes or chromosome sets, called aneuploidy, may be lethal or may give rise to genetic disorders. Genetic counseling is offered for families that may carry a chromosome rearrangement.

The gain or loss of DNA from chromosomes can lead to a variety of genetic disorders. Human examples include:

- Cri du chat, which is caused by the deletion of part of the short arm of chromosome 5. "Cri du chat" means "cry of the cat" in French; the condition was so-named because affected babies make high-pitched cries that sound like those of a cat. Affected individuals have wide-set eyes, a small head and jaw, moderate to severe mental health problems, and are very short.

- Down syndrome, the most common trisomy, usually caused by an extra copy of chromosome 21 (trisomy 21). Characteristics include decreased muscle tone, stockier build, asymmetrical skull, slanting eyes and mild to moderate developmental disability.

- Edwards syndrome, or trisomy-18, the second most common trisomy. Symptoms include motor retardation, developmental disability and numerous congenital anomalies causing serious health problems. Ninety percent of those affected die in infancy. They have characteristic clenched hands and overlapping fingers.

- Isodicentric 15, also called idic(15), partial tetrasomy 15q, or inverted duplication 15 (inv dup 15).

- Jacobsen syndrome, which is very rare. It is also called the terminal 11q deletion disorder. Those affected have normal intelligence or mild developmental disability, with poor expressive language skills. Most have a bleeding disorder called Paris-Trousseau syndrome.

- Klinefelter syndrome (XXY). Men with Klinefelter syndrome are usually sterile and tend to be taller and have longer arms and legs than their peers. Boys with the syndrome are often shy and quiet and have a higher incidence of speech delay and dyslexia. Without testosterone treatment, some may develop gynecomastia during puberty.

- Patau Syndrome, also called D-Syndrome or trisomy-13. Symptoms are somewhat similar to those of trisomy-18, without the characteristic folded hand.

- Small supernumerary marker chromosome. This means there is an extra, abnormal chromosome. Features depend on the origin of the extra genetic material. Cat-eye syndrome and isodicentric chromosome 15 syndrome (or Idic15) are both caused by a supernumerary marker chromosome, as is Pallister–Killian syndrome.

- Triple-X syndrome (XXX). XXX girls tend to be tall and thin and have a higher incidence of dyslexia.

- Turner syndrome (X instead of XX or XY). In Turner syndrome, female sexual characteristics are present but underdeveloped. Females with Turner syndrome often have a short stature, low hairline, abnormal eye features and bone development and a "caved-in" appearance to the chest.

- Wolf–Hirschhorn syndrome, which is caused by partial deletion of the short arm of chromosome 4. It is characterized by growth retardation, delayed motor skills development, "Greek Helmet" facial features, and mild to profound mental health problems.

- XYY syndrome. XYY boys are usually taller than their siblings. Like XXY boys and XXX girls, they are more likely to have learning difficulties.

Sperm Aneuploidy

Exposure of males to certain lifestyle, environmental and/or occupational hazards may increase the risk of aneuploid spermatozoa. In particular, risk of aneuploidy is increased by tobacco smoking, and occupational exposure to benzene, insecticides, and perfluorinated compounds. Increased aneuploidy is often associated with increased DNA damage in spermatozoa.

Gene

Gene is the unit of hereditary information that occupies a fixed position (locus) on a chromosome. Genes achieve their effects by directing the synthesis of proteins.

In eukaryotes (such as animals, plants, and fungi), genes are contained within the cell nucleus. The mitochondria (in animals) and the chloroplasts (in plants) also contain small subsets of genes distinct from the genes found in the nucleus. In prokaryotes (organisms lacking a distinct nucleus, such as bacteria), genes are contained in a single chromosome that is free-floating in the cell cytoplasm. Many bacteria also contain plasmids—extrachromosomal genetic elements with a small number of genes.

The number of genes in an organism's genome (the entire set of chromosomes) varies significantly between species. For example, whereas the human genome contains an estimated 20,000 to 25,000 genes, the genome of the bacterium Escherichia coli O157:H7 houses precisely 5,416 genes. Arabidopsis thaliana—the first plant for which a complete genomic sequence was recovered—has roughly 25,500 genes; its genome is one of the smallest known to plants. Among extant independently replicating organisms, the bacterium Mycoplasma genitalium has the fewest number of genes, just 517.

Chemical Structure of Genes

Genes are composed of deoxyribonucleic acid (DNA), except in some viruses, which have genes consisting of a closely related compound called ribonucleic acid (RNA). A DNA molecule is composed of two chains of nucleotides that wind about each other to resemble a twisted ladder. The sides of the ladder are made up of sugars and phosphates, and the rungs are formed by bonded pairs of nitrogenous bases. These bases are adenine (A), guanine (G), cytosine (C), and thymine (T). An A on one chain bonds to a T on the other (thus forming an A–T ladder rung); similarly, a C on one chain bonds to a G on the other. If the bonds between the bases are broken, the two chains unwind, and free nucleotides within the cell attach themselves to the exposed bases of the now-separated chains. The free nucleotides line up along each chain according to the base-pairing rule—A bonds to T, C bonds to G. This process results in the creation of two identical DNA molecules from one original and is the method by which hereditary information is passed from one generation of cells to the next.

Gene Transcription And Translation

The sequence of bases along a strand of DNA determines the genetic code. When the product of a particular gene is needed, the portion of the DNA molecule that contains that gene will split. Through the process of transcription, a strand of RNA with bases complementary to those of the gene is created from the free nucleotides in the cell. (RNA has the base uracil [U] instead of thymine, so A and U form base pairs during RNA synthesis.) This single chain of RNA, called messenger RNA (mRNA), then passes to the organelles called ribosomes, where the process of translation, or protein synthesis, takes place. During translation, a second type of RNA, transfer RNA (tRNA), matches up the nucleotides on mRNA with specific amino acids. Each set of three nucleotides codes for one amino acid. The series of amino acids built according to the sequence of nucleotides forms a polypeptide chain; all proteins are made from one or more linked polypeptide chains.

Experiments conducted in the 1940s indicated one gene being responsible for the assembly of one enzyme, or one polypeptide chain. This is known as the one gene–one enzyme hypothesis. However, since this discovery, it has been realized that not all genes encode an enzyme and that some enzymes are made up of several short polypeptides encoded by two or more genes.

Human Reproduction

Human reproduction is essential for the continuance of the human species. Humans reproduce sexually by the uniting of the female and male sex cells. Although the reproductive systems of the male and female are different, they are structured to function together to achieve internal fertilization.

It is a characteristic of all living things on Earth that they reproduce or produce offspring, and humans are no different. If humans are examined as large, complex land mammals, then we can say from a strictly biological point of view that the male and female have the same role as other mammals. The male's job is to produce sperm cells and deliver them into the female reproductive tract. The female's job is to produce ova (eggs), receive the sperm, and nourish the embryo that grows inside her. She must also give birth and produce milk for the offspring during its early years.

Human Reproductive System

Both the male and female reproductive structures have 3 levels of organisation:

1. Production of sex cells.

2. Transport tubes.

3. Glands to secrete hormones.

The Male Reproductive System

The testes produce the sperm cells by meiosis. The temperature must be lower than body temperature for this to occur. There are tubules that are lines with sperm producing cells. Testosterone, the male sex hormone, is also produced in the testes. Once the sperm are produced they mature in the epidymis. This structure is located outside of the testis. If they are not released within about 6 weeks they are broken down and released to the bloodstream by a process called resorption. The sperm are carried to the urethra by the sperm duct. The urethra carries both sperm and urine.

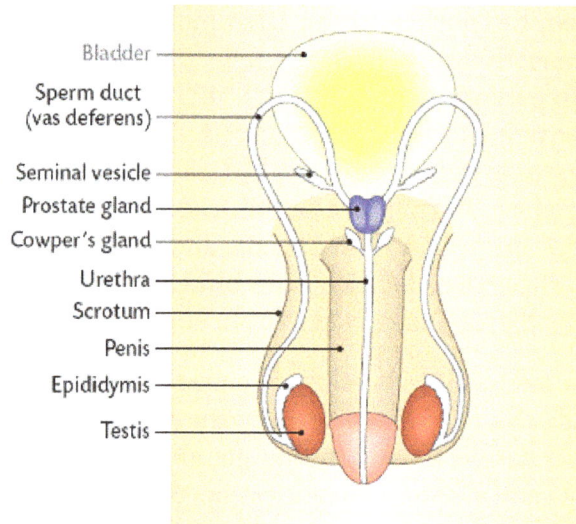

The gonad is the name for the organ that produces sex cells in organs. The male gonads are called the testes. The testes are contained in the scrotum.

The sperm cells are carried within a liquid called semen. The semen is produced by the seminal vescicles, the prostate gland, and Cowper's glands. The semen also contains nourishment for the sperm cells.

Sperm cells are released by ejaculation. About 50-300 million sperm cells are released at one time.

Sperm cells, also called spermatozoa, are haploid containing 23 chromosomes. Their production begins at puberty.

The penis is adapted to place sperm cells into the female. The tip is called the glans. Erection occurs when blood rushes into the penis.

The Structure of the Sperm Cell

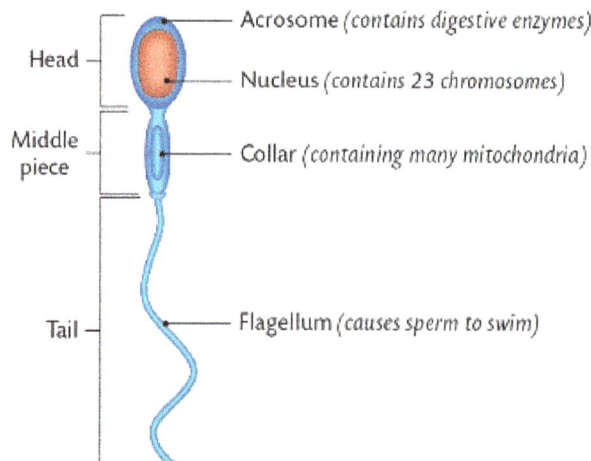

Male Hormones

Male hormones are produced by the pituitary gland during puberty. They are:

FSH- Follicle Stimulating Hormone: This causes the production of sperm by meiosis.

LH- Leuteinising Hormone: Stimulates the testes to produce testosterone.

Testosterone

During the period of pregnancy testosterone causes the development of primary male sex characteristics. These include the development of the penis and the other male reproductive parts.

Later in life, at puberty, testosterone causes the enlargement of the reproductive parts as well as the development of secondary sexual characteristics. These are characteristics that distinguish males from females.

Male secondary sexual characteristics included:

1. Hair growth on the face, underarm, chest and pubic region.

2. Enlarged larynx producing a deeper voice.

3. Wider shoulders.

4. Greater skeletal muscular development.

5. Growth in height and weight.

Male Infertility

The most common cause of male infertility is the low production of sperm. There are many causes of low sperm production. Stress, alcohol and drug abuse, high temperature of the testes, and low testosterone production are all causes.

The Female Reproductive System

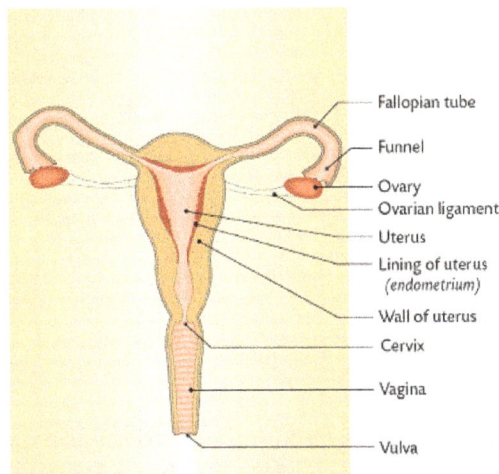

Labels: Fallopian tube, Funnel, Ovary, Ovarian ligament, Uterus, Lining of uterus (endometrium), Wall of uterus, Cervix, Vagina, Vulva

Ovaries

Ovaries produce eggs and female hormones. At puberty there are about 40,000 diploid eggs. Each egg is enclosed in a group of cells called a follicle. About 20 haploid eggs are produced each month. Usually all but one die. The haploid egg cell is called the ovum and is surrounded by the Graafian follicle which produces the female hormone called oestrogen. Ovulation is the release of the egg from the follicle. This occurs when the follicle bursts.

Fallopian Tubes

The fallopian tubes are about 12 cm long and have ends that are funnel shaped. These ends collect the egg after ovulation. Cilia and peristalsis move the egg along the tube. The egg will die in the tube if it is not fertilised.

Uterus

The uterus, also known as the womb, is made of involuntary muscle. It is lines with the endometrium. This lining thickens with cells and blood every month. This happens in order to nourish the embryo (if present). The opening of the uterus is called the cervix.

Vagina

The vagina is a muscular tube which allows the sperm to enter the female as well as the baby to exit. It is lined with mucous secreting cells. The urethra opens near the vagina. The vagina is protected by folds of skin called the vulva. The hymen partially blocks the entrance of the vagina. It is broken by sexual intercourse or with the use of tampons.

The Menstrual Cycle

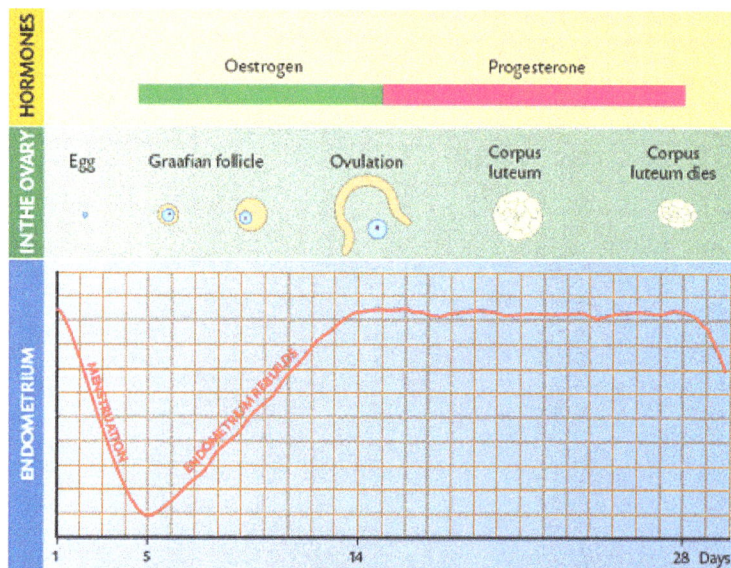

The menstrual cycle occurs every 28 days from puberty to menopause (the end of the females reproductive life). It occurs only if fertilisation of the egg has not taken place.

The typical events of the menstrual cycle are:

Day 1 to day 5-

    a. The endometrium breaks down and is shed from the body. This is called menstruation.

    b. Meiosis occus in the ovary to produce a new egg surrounded by the Graafian follicle.

Day 6 to day 13-

    a. Oestrogen is produced by the Graafian follicle. Oestrogen also stimulates the endometrium to thicken again. One Graafian follicle with one egg develops.

    b. Oestrogen stimulates the production of LH (leuteinising hormone)

Day 14-

    a. The surge of LH stimulates ovulation.

    b. The egg enters the funnel of the Fallopian tube. It can be fertilised for the next 48 hours.

Day 15 to day 26-

    a. The corpus luteum (yellow body) develops from the remains of the Graafian follicle. This produced progesterone and some oestrogen. The progesterone causes the endometrium to continue to thicken. It also prevents new eggs from forming.

    b. The egg that was released at day 14 will die if it is not fertilised.

    c. If fertilisation did not take place the corpus luteum begins to degenerate.

Day 26 to day 28-

    a. Oestrogen and progesterone levels decline.

    b. The endometrium begins to break down.

    c. Day one of the cycle begins.

Female Hormones

In summary:

Endometrium thickened by oestrogen in days 1-14 and by progesterone in days 15-28.

Both prevent egg development.

At puberty, oestrogen causes the primary female sexual characteristics of the growth of the sex organs. At puberty both oestrogen and progesterone cause the secondary female characteristics.

They include:

    a. The enlargement of the breasts

    b. Widening of the hips

    c. Increased body fat

d. Growth of public and underarm hair

e. General growth spurt in height

Hormone	Site of Production	Time of Production	Functions
FSH-follicle stimulating hormone	Pituitary Gland	Days 1-5 of menstrual cycle	Stimulates egg production within Graafian follicles. Sometimes used in fertility treatment to stimulate egg production. Graafian follicles secrete oestrogen.
Oestrogen	Graafian follicle	Days 5-14 of menstrual cycle	Development of endometrium. Inhibits FSH so no new eggs develop. Stimulates the release of LH (luteinising hormone).
LH- leuteinising hormone	Pituitary Gland	Day 14 of menstrual cycle	Causes ovulation. Causes Graafian follicle to develop into the corpus luteum. The corpus luteum makes progesterone.
Progesterone	Corpus luteum	Days 14-28 of menstrual cycle	Maintains endometrium. Inhibits FSH so no new eggs develop. Inhibits LH so no new ovulations occur. Prevents contractions of the uterus.

Female Infertility

Female infertility is the inability to conceive either by fertilisation failure or implantation failure. Egg cell formation or ovulation may not occur due to a hormone imbalance. The egg cell may not be able pass to the uterus due to blockage of the Fallopian tubes. Treatment with hormones may be successful. In-vitro fertilisation and implantation is often used to treat female infertility.

Fibroids

Fibroids are benign tumours of the uterus. They are slow growing and range in size. Small fibroids produce no symptoms while large ones can cause heavy and prolonged menstrual bleeding. They can also cause pain, miscarriage, or infertility. Some science shows that they may be caused as an abnormal response to oestrogen. Large fibroids are removed by surgery. In severe cases where there are many large fibroids the uterus may have to be removed. This is called a hysterectomy.

Copulation

Copulation is also called coitus or sexual intercourse. During this process the penis moves into the vagina in order to deposit semen which contains sperm cells. The depositing of the semen is called insemination.

Fertilisation

a.   After insemination the sperm will move up the Fallopian tubes.

b.   If ovulation has occurred and an egg is present the egg will release a chemical that attracts the sperm. This is called chemotaxix.

c.   The sperm that reaches the egg will use an enzyme in its acrosomes to make an opening in the membrane of the egg.

d.   Once one sperm enters the egg (only the head enters) the egg forms a membrane that prevents other sperm from entering.

e.   The nucleus of the egg fuses with the nucleus of the egg. A diploid zygote forms.

f.   Fertilisation may take place during days 11-16 of the menstrual cycle.

Events of Fertilisation

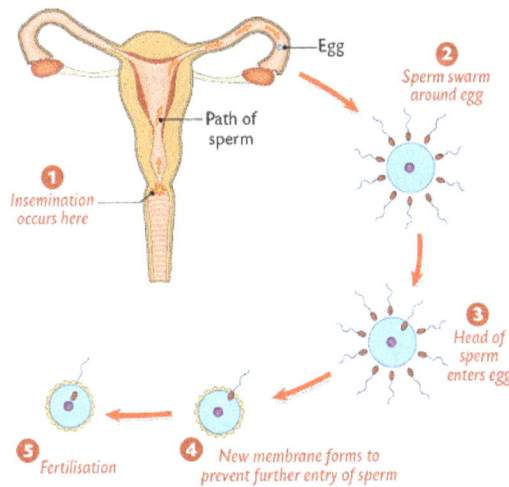

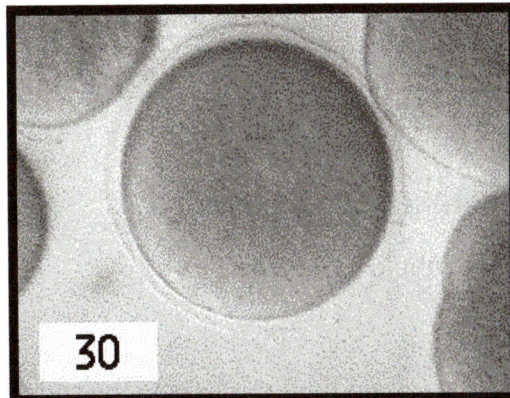

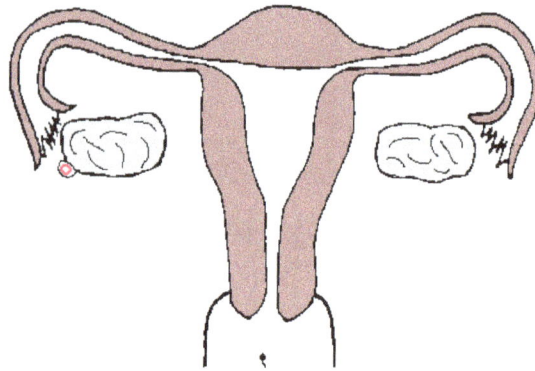

IMPLANTATION

About 6-9 days after fertilisation the fertilised egg becomes embedded into the lining of the uterus. The zygote has now become an embryo. A membrane called the amnion develops around the ebbryo. This membrane will secrete amnion fluid which surrounds and protects the embryo.

Placenta Formation

a. After implantation the embryo forms another membrane called the chorion. This surrounds the embryo.

b. Projections of the chorion called villi join with blood vessels in the endometrium to form the placenta.

c. The placenta become fully functional in about 3 months.

d. The umbilical cord connects the embryo (at the navel) with the placenta.

Functions of the Placenta

1. Protection of the Embryo

a. It hinders the entry of pathogens from the mother.

b. It allows the entry of antibodies from the mother (passive induced immunity).

c. It keeps the embryo separated from the mothers higher blood pressure.

d. It prevents exchange of red blood cells avoiding the deadly possibility of agglutination

2. Gas Exchange

a. It supplies O_2 from the mother.

b. It excretes CO_2 from the embryo to the mothers blood.

3. Nutrient Supply

Glucose, amino acids, lipids, vitamins and minerals pass to the embryo from the mothers blood.

4. Endocrine

It secretes a variety of hormones including oestrogen and progesterone. The hormones maintain the pregnancy and prepare the mothers body for birth and lactation.

5. Excretion

Metabolic wastes, CO2 and urea, pass from embryo into the mothers blood.

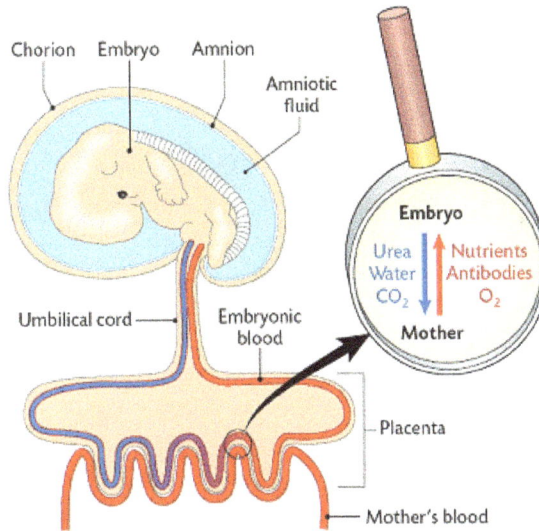

Early Developmet of the Zygote

1. The zygote divides many times by cleavage (increase in the number of cells by division but no overall increase in size) to double its cell number. A solid clump of about 100 cells called the morula is formed.

2. About 5 days after fertilisation the morula develops into a hollow ball called a blastocyst. The outer cells of the blastocyst form the trophoplast. This will become the membranes around the embryo. The inner cells, called the inner mass will become the embryo.

3. The blastocyst is pushed down the fallopian tube and into the uterus for implantation.

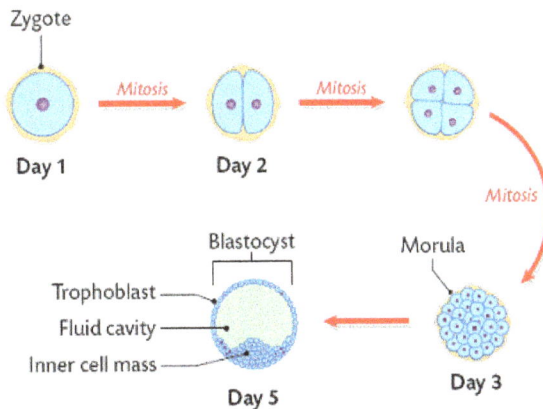

In normal development the sperm fertilizes the egg. The cortical reaction occurs raising the fertilization membrane and cell divisions occur until the blastula stage. When the embryo reaches the blastula stage the embryo releases an enzyme that dissolves the fertilization membrane and the young embryo swims free to continue development.

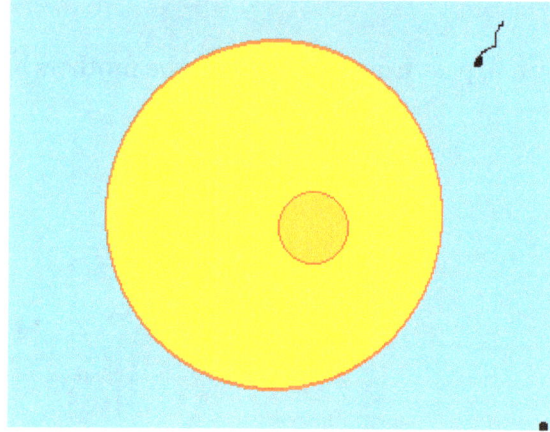

Developmet of The Embryo

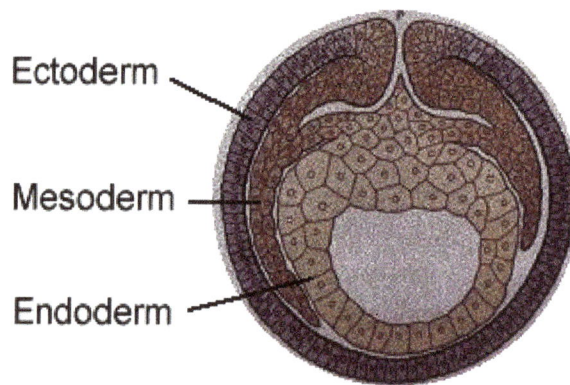

The mesoderm cells further develop into the muscles and blood, the endoderm develops into the digestive tract and lungs, and the ectoderm develops into the skin, nerves and brain.

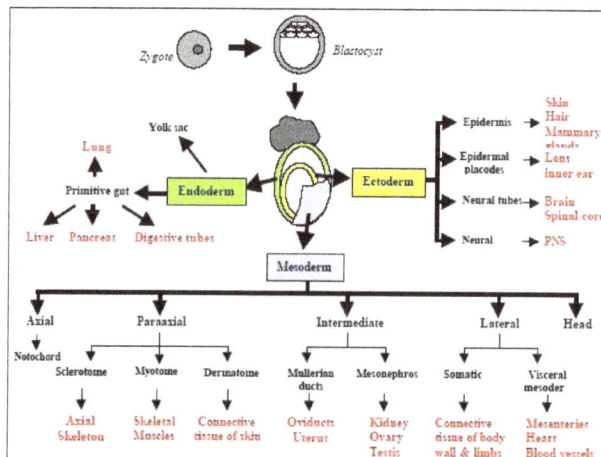

Week to Week Development

4 Weeks after fertilisation	1. Heart formsand starts to beat.
	2. Brain develops
	3. Umbilical cord forms.
5 Weeks after ferlilisation	1. Internal organs start to form.
	2. Limbs start to form.
	(Embryo highly vulnerable to alcohol and drugs)
6 Weeks after ferlilisation	1. Eyes become visible.
	2. Mouth, nose and ears begin to form.
8 Weeks after ferlilisation	1. Tailis gone.
	2. Facebecomes human looking.
	3. Majororgans are formed.
	4. Ovariesor testes are seen.
	5. Bone replaces cartilage.
	6. The embryo is now called afoetus.
12 Weeks afterfertilisation	1. Bone growth continues to replace cartilage.
	2. Nerves and muscles coordinate arm and leg movement.
	3. Thumb sucking and kicking begin.
	4. Milk (baby) teeth form.
	5. Foetus takes amniotic fluid into mouth and releases urine and faeces into amniotic fluid.
	6. External sex organs clearly seen. Gender can be determined with a scan.

The Gestation period is the length of time from fertilisation to birth. In humans it is generally 266 days (38 weeks/9months).

Birth

1. The placenta stops producing progesterone. This causes the walls of the uterus to contract.

2. Oxytocin is produced by the pituitary gland. This hormone causes contractions of the uterine muscle. This is the beginning of labour.

3. There are 3 stages of labour:

    a. Stage 1: The contractions push the foetus down toward the cervix. The membrane around the foetus (amnion) breaks. The amniotic fluid is released through the vagina.

    b. Stage 2: The cervix dilates (widens) and the foetus is pushed out through the cervix and vagina. At this time the umbilical cord is cut.

    c. Stage 3: The placenta and foetal membrane (afterbirth) are released through the vagina.

Lactation

Lactation is the secretion of milk by the mammary glands of the mother. Colostrum is a thick yellow fluid produced during the first few days. It is low in fat and sugar but rich in minerals, protein, and antibodies.

Prolactin is a hormone produced by the pituitary gland. This hormone stimulates milk production. The suckling of a baby at the breast stimulates the mothers pituitary to release prolactin. When breast feeding stops the mother stops secreting prolactin and therefore stops producing milk. Suckling also stimulates the pituitary to secrete oxytocin. Oxytocin causes the milk ducts to contract ejecting the milk from the breast.

Benefits of Breast Feeding

Human milk has a lot of advantages for the babys growth and development. Human milk is nutritionally balanced for a developing human baby. It also contains a wide variety of beneficial chemicals that include mothers antibodies. These antibodies protect the child against common pathogens. Human milk also contains chemicals favourable for brain growth and development. Human milk also encourages the growth of mutualistic bacteria in the large intestine.

References

- Berg JM, Tymoczko JL, Stryer L (2002). Biochemistry (5th ed.). WH Freeman and Company. pp. 118–19, 781–808. ISBN 0-7167-4684-0. OCLC 179705944

- Tinoco I & Bustamante C (October 1999). "How RNA folds". Journal of Molecular Biology. 293 (2): 271–81. doi:10.1006/jmbi.1999.3001. PMID 10550208

- Marlaire R (3 March 2015). "NASA Ames Reproduces the Building Blocks of Life in Laboratory". NASA. Retrieved 5 March 2015

- Kiss T (July 2001). "Small nucleolar RNA-guided post-transcriptional modification of cellular RNAs". The EMBO Journal. 20 (14): 3617–22. doi:10.1093/emboj/20.14.3617. PMC 125535. PMID 11447102

- Barciszewski J, Frederic B, Clark C (1999). RNA biochemistry and biotechnology. Springer. pp. 73–87. ISBN 0-7923-5862-7. OCLC 52403776

- Higgs PG (August 2000). "RNA secondary structure: physical and computational aspects". Quarterly Reviews of Biophysics. 33 (3): 199–253. doi:10.1017/S0033583500003620. PMID 11191843

- Designation of the two strands of DNA Archived 24 April 2008 at the Wayback Machine. JCBN/NC-IUB Newsletter 1989. Retrieved 7 May 2008

- Johnson TB, Coghill RD (1925). "Pyrimidines. CIII. The discovery of 5-methylcytosine in tuberculinic acid, the nucleic acid of the tubercle bacillus". Journal of the American Chemical Society. 47: 2838–44

- Jankowski JA, Polak JM (1996). Clinical gene analysis and manipulation: Tools, techniques and troubleshooting. Cambridge University Press. p. 14. ISBN 0-521-47896-0. OCLC 33838261

Chapter 4

Human Skeletal System

The human skeleton is the system that provides the framework to the human body and consists of around 206 bones in adulthood. The human skeleton can be divided into two groups, the axial skeleton and the appendicular skeleton, which have elaborately covered in this chapter. It further elucidates the constituting structures of the skeletal systems, such as shoulder girdle, pelvis, thighs, ankles, human head, etc.

Human skeletal system, the internal skeleton that serves as a framework for the body. This framework consists of many individual bones and cartilages. There also are bands of fibrous connective tissue—the ligaments and the tendons—in intimate relationship with the parts of the skeleton.

The human skeleton, like that of other vertebrates, consists of two principal subdivisions, each with origins distinct from the others and each presenting certain individual features. These are (1) the axial, comprising the vertebral column—the spine—and much of the skull, and (2) the appendicular, to which the pelvic (hip) and pectoral (shoulder) girdles and the bones and cartilages of the limbs belong.

When one considers the relation of these subdivisions of the skeleton to the soft parts of the human body—such as the nervous system, the digestive system, the respiratory system, the cardiovascular system, and the voluntary muscles of the muscle system—it is clear that the functions of the skeleton are of three different types: support, protection, and motion. Of these functions, support is the most primitive and the oldest; likewise, the axial part of the skeleton was the first to evolve. The vertebral column, corresponding to the notochord in lower organisms, is the main support of the trunk.

The central nervous system lies largely within the axial skeleton, the brain being well protected by the cranium and the spinal cord by the vertebral column, by means of the bony neural arches (the arches of bone that encircle the spinal cord) and the intervening ligaments.

A distinctive characteristic of humans as compared with other mammals is erect posture. The human body is to some extent like a walking tower that moves on pillars, represented by the legs. Tremendous advantages have been gained from this erect posture, the chief among which has been the freeing of the arms for a great variety of uses. Nevertheless, erect posture has created a number of mechanical problems—in particular, weight bearing. These problems have had to be met by adaptations of the skeletal system.

Protection of the heart, lungs, and other organs and structures in the chest creates a problem somewhat different from that of the central nervous system. These organs, the function of which involves motion, expansion, and contraction, must have a flexible and elastic protective covering.

Such a covering is provided by the bony thoracic basket, or rib cage, which forms the skeleton of the wall of the chest, or thorax. The connection of the ribs to the breastbone—the sternum—is in all cases a secondary one, brought about by the relatively pliable rib (costal) cartilages. The small joints between the ribs and the vertebrae permit a gliding motion of the ribs on the vertebrae during breathing and other activities. The motion is limited by the ligamentous attachments between ribs and vertebrae.

The third general function of the skeleton is that of motion. The great majority of the skeletal muscles are firmly anchored to the skeleton, usually to at least two bones and in some cases to many bones. Thus, the motions of the body and its parts, all the way from the lunge of the football player to the delicate manipulations of a handicraft artist or of the use of complicated instruments by a scientist, are made possible by separate and individual engineering arrangements between muscle and bone.

Axial and Visceral Skeleton

The Cranium

The cranium—the part of the skull that encloses the brain—is sometimes called the braincase, but its intimate relation to the sense organs for sight, sound, smell, and taste and to other structures makes such a designation somewhat misleading.

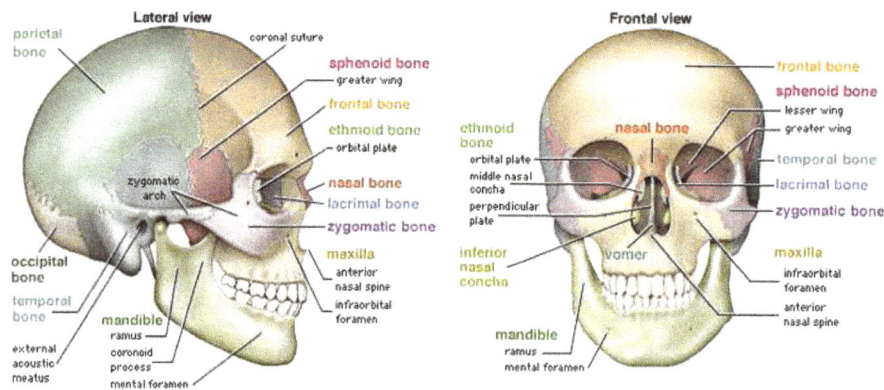

(Left) Lateral and (right) frontal views of the human skull

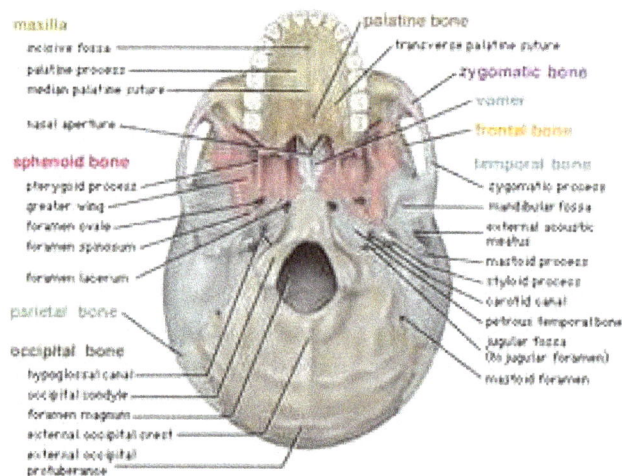

Inferior view of the human skull

Development of Cranial Bones

The cranium is formed of bones of two different types of developmental origin—the cartilaginous, or substitution, bones, which replace cartilages preformed in the general shape of the bone; and membrane bones, which are laid down within layers of connective tissue. For the most part, the substitution bones form the floor of the cranium, while membrane bones form the sides and roof.

The range in the capacity of the cranial cavity is wide but is not directly proportional to the size of the skull, because there are variations also in the thickness of the bones and in the size of the air pockets, or sinuses. The cranial cavity has a rough, uneven floor, but its landmarks and details of structure generally are consistent from one skull to another.

The cranium forms all the upper portion of the skull, with the bones of the face situated beneath its forward part. It consists of a relatively few large bones, the frontal bone, the sphenoid bone, two temporal bones, two parietal bones, and the occipital bone. The frontal bone underlies the forehead region and extends back to the coronal suture, an arching line that separates the frontal bone from the two parietal bones, on the sides of the cranium. In front, the frontal bone forms a joint with the two small bones of the bridge of the nose and with the zygomatic bone (which forms part of the cheekbone; *see below* The facial bones and their complex functions), the sphenoid, and the maxillary bones. Between the nasal and zygomatic bones, the horizontal portion of the frontal bone extends back to form a part of the roof of the eye socket, or orbit; it thus serves an important protective function for the eye and its accessory structures.

Each parietal bone has a generally four-sided outline. Together they form a large portion of the side walls of the cranium. Each adjoins the frontal, the sphenoid, the temporal, and the occipital bones and its fellow of the opposite side. They are almost exclusively cranial bones, having less relation to other structures than the other bones that help to form the cranium.

Functions

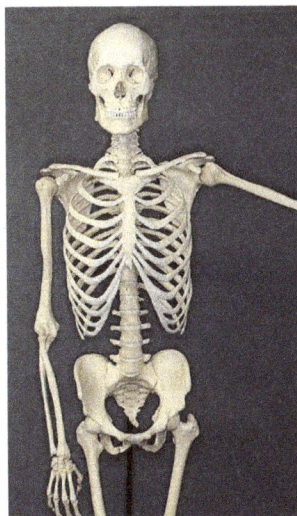

A human skeleton on exhibit at the Museum of Osteology, Oklahoma City, Oklahoma

The skeleton serves six major functions: support, movement, protection, production of blood cells, storage of minerals and endocrine regulation.

Support

The skeleton provides the framework which supports the body and maintains its shape. The pelvis, associated ligaments and muscles provide a floor for the pelvic structures. Without the rib cages, costal cartilages, and intercostal muscles, the lungs would collapse.

Movement

The joints between bones allow movement, some allowing a wider range of movement than others, e.g. the ball and socket joint allows a greater range of movement than the pivot joint at the neck. Movement is powered by skeletal muscles, which are attached to the skeleton at various sites on bones. Muscles, bones, and joints provide the principal mechanics for movement, all coordinated by the nervous system.

It is believed that the reduction of human bone density in prehistoric times reduced the agility and dexterity of human movement. Shifting from hunting to agriculture has caused human bone density to reduce significantly.

Protection

The skeleton helps to protect our many vital internal organs from being damaged.

- The skull protects the brain

- The vertebrae protect the spinal cord.

- The rib cage, spine, and sternum protect the lungs, heart and major blood vessels.

Blood Cell Production

The skeleton is the site of haematopoiesis, the development of blood cells that takes place in the bone marrow. In children, haematopoiesis occurs primarily in the marrow of the long bones such as the femur and tibia. In adults, it occurs mainly in the pelvis, cranium, vertebrae, and sternum.

Storage

The bone matrix can store calcium and is involved in calcium metabolism, and bone marrow can store iron in ferritin and is involved in iron metabolism. However, bones are not entirely made of calcium, but a mixture of chondroitin sulfate and hydroxyapatite, the latter making up 70% of a bone. Hydroxyapatite is in turn composed of 39.8% of calcium, 41.4% of oxygen, 18.5% of phosphorus, and 0.2% of hydrogen by mass. Chondroitin sulfate is a sugar made up primarily of oxygen and carbon.

Endocrine Regulation

Bone cells release a hormone called osteocalcin, which contributes to the regulation of blood sugar (glucose) and fat deposition. Osteocalcin increases both the insulin secretion and sensitivity, in addition to boosting the number of insulin-producing cells and reducing stores of fat.

Gender Differences

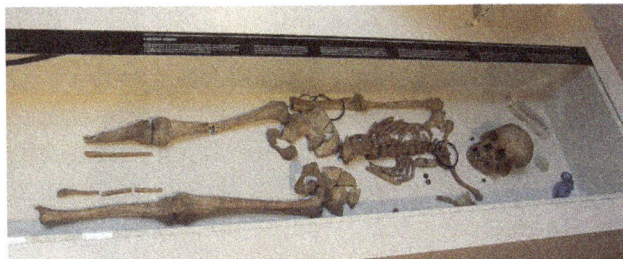

During construction of the York to Scarborough Railway Bridge, workmen discovered a large stone coffin, close to the River Ouse. Inside was a skeleton, accompanied by an array of unusual and expensive objects. Study of the skeleton has revealed that it belonged to a woman

Anatomical differences between human males and females are highly pronounced in some soft tissue areas, but tend to be limited in the skeleton. The human skeleton is not as sexually dimorphic as that of many other primate species, but subtle differences between sexes in the morphology of the skull, dentition, long bones, and pelvis are exhibited across human populations. In general, female skeletal elements tend to be smaller and less robust than corresponding male elements within a given population. It is not known whether or to what extent those differences are genetic or environmental.

Skull

A variety of gross morphological traits of the human skull demonstrate sexual dimorphism, such as the median nuchal line, mastoid processes, supraorbital margin, supraorbital ridge, and the chin.

Dentition

Human inter-sex dental dimorphism centers on the canine teeth, but it is not nearly as pronounced as in the other great apes.

Long Bones

Long bones are generally larger in males than in females within a given population. Muscle attachment sites on long bones are often more robust in males than in females, reflecting a difference in overall muscle mass and development between sexes. Sexual dimorphism in the long bones is commonly characterized by morphometric or gross morphological analyses.

Pelvis

The human pelvis exhibits greater sexual dimorphism than other bones, specifically in the size and shape of the pelvic cavity, ilia, greater sciatic notches, and the sub-pubic angle. The Phenice method is commonly used to determine the sex of an unidentified human skeleton by anthropologists with 96% to 100% accuracy in some populations.

Women's pelvises are wider in the pelvic inlet and are wider throughout the pelvis to allow for child birth. The sacrum in the women's pelvis is curved inwards to allow the child to have a "funnel" to assist in the child's pathway from the uterus to the birth canal.

Axial Skeleton

The axial skeleton forms the central axis of the body and includes the bones of the skull, ossicles of the middle ear, hyoid bone of the throat, vertebral column, and the thoracic cage (ribcage) (Figure below). The function of the axial skeleton is to provide support and protection for the brain, the spinal cord, and the organs in the ventral body cavity. It provides a surface for the attachment of muscles that move the head, neck, and trunk, performs respiratory movements, and stabilizes parts of the appendicular skeleton.

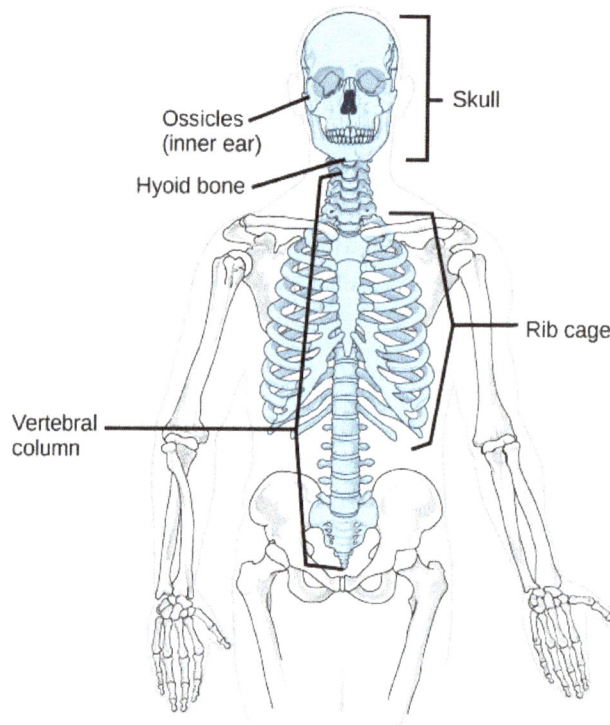

Figure: The axial skeleton consists of the bones of the skull, ossicles of the middle ear, hyoid bone, vertebral column, and rib cage

The Skull

The bones of the skull support the structures of the face and protect the brain. The skull consists of 22 bones, which are divided into two categories: cranial bones and facial bones. The cranial bones are eight bones that form the cranial cavity, which encloses the brain and serves as an attachment site for the muscles of the head and neck. The eight cranial bones are the frontal bone, two parietal bones, two temporal bones, occipital bone, sphenoid bone, and the ethmoid bone. Although the bones developed separately in the embryo and fetus, in the adult, they are tightly fused with connective tissue and adjoining bones do not move (Figure below).

The auditory ossicles of the middle ear transmit sounds from the air as vibrations to the fluid-filled cochlea. The auditory ossicles consist of six bones: two malleus bones, two incus bones, and two stapes (one of each bone on each side). These are the smallest bones in the body and are unique to mammals.

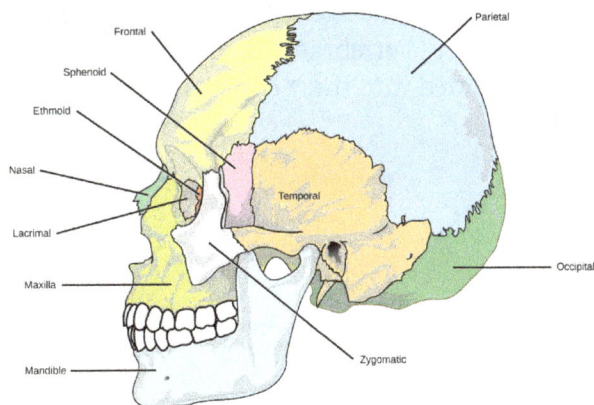

Figure: The bones of the skull support the structures of the face and protect the brain

Fourteen facial bones form the face, provide cavities for the sense organs (eyes, mouth, and nose), protect the entrances to the digestive and respiratory tracts, and serve as attachment points for facial muscles. The 14 facial bones are the nasal bones, the maxillary bones, zygomatic bones, palatine, vomer, lacrimal bones, the inferior nasal conchae, and the mandible. All of these bones occur in pairs except for the mandible and the vomer (Figure below).

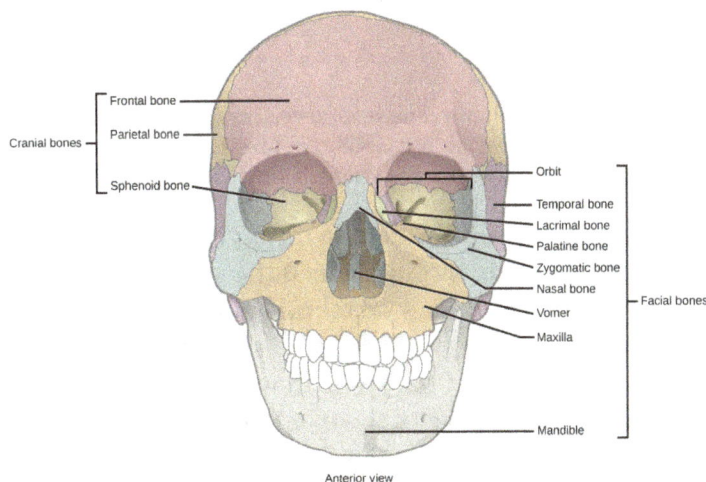

Figure: The cranial bones, including the frontal, parietal, and sphenoid bones, cover the top of the head. The facial bones of the skull form the face and provide cavities for the eyes, nose, and mouth

Although it is not found in the skull, the hyoid bone is considered a component of the axial skeleton. The hyoid bone lies below the mandible in the front of the neck. It acts as a movable base for the tongue and is connected to muscles of the jaw, larynx, and tongue. The mandible articulates with the base of the skull. The mandible controls the opening to the airway and gut. In animals with teeth, the mandible brings the surfaces of the teeth in contact with the maxillary teeth.

The Vertebral Column

The vertebral column, or spinal column, surrounds and protects the spinal cord, supports the head, and acts as an attachment point for the ribs and muscles of the back and neck. The adult vertebral column comprises 26 bones: the 24 vertebrae, the sacrum, and the coccyx bones. In the adult, the sacrum is typically composed of five vertebrae that fuse into one. The coccyx is typically

3–4 vertebrae that fuse into one. Around the age of 70, the sacrum and the coccyx may fuse together. We begin life with approximately 33 vertebrae, but as we grow, several vertebrae fuse together. The adult vertebrae are further divided into the 7 cervical vertebrae, 12 thoracic vertebrae, and 5 lumbar vertebrae (Figure below).

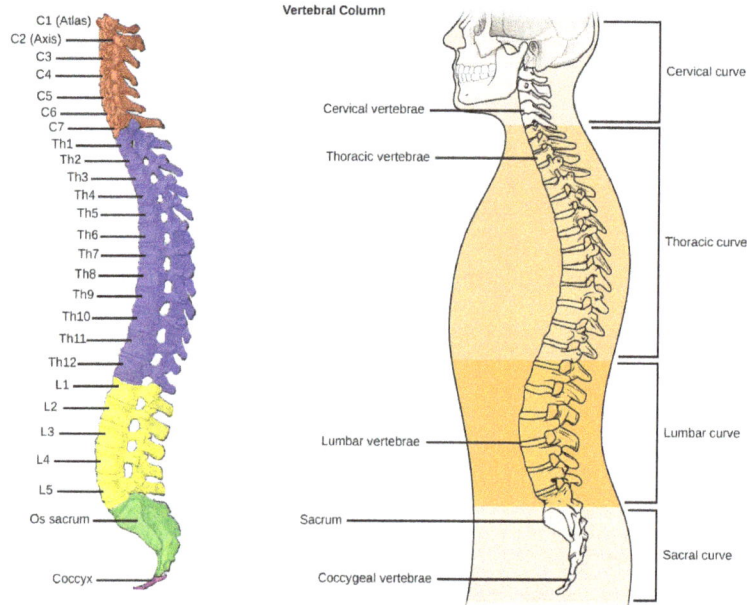

Figure: (a) The vertebral column consists of seven cervical vertebrae (C1–7) twelve thoracic vertebrae (Th1–12), five lumbar vertebrae (L1–5), the os sacrum, and the coccyx. (b) Spinal curves increase the strength and flexibility of the spine

Each vertebral body has a large hole in the center through which the nerves of the spinal cord pass. There is also a notch on each side through which the spinal nerves, which serve the body at that level, can exit from the spinal cord. The vertebral column is approximately 71 cm (28 inches) in adult male humans and is curved, which can be seen from a side view. The names of the spinal curves correspond to the region of the spine in which they occur. The thoracic and sacral curves are concave (curve inwards relative to the front of the body) and the cervical and lumbar curves are convex (curve outwards relative to the front of the body). The arched curvature of the vertebral column increases its strength and flexibility, allowing it to absorb shocks like a spring (Figure above).

Intervertebral discs composed of fibrous cartilage lie between adjacent vertebral bodies from the second cervical vertebra to the sacrum. Each disc is part of a joint that allows for some movement of the spine and acts as a cushion to absorb shocks from movements such as walking and running. Intervertebral discs also act as ligaments to bind vertebrae together. The inner part of discs, the nucleus pulposus, hardens as people age and becomes less elastic. This loss of elasticity diminishes its ability to absorb shocks.

The Thoracic Cage

The thoracic cage, also known as the ribcage, is the skeleton of the chest, and consists of the ribs, sternum, thoracic vertebrae, and costal cartilages (Figure below). The thoracic cage encloses and protects the organs of the thoracic cavity, including the heart and lungs. It also provides support for the shoulder girdles and upper limbs, and serves as the attachment point for the diaphragm, muscles of the back, chest, neck, and shoulders. Changes in the volume of the thorax enable breathing.

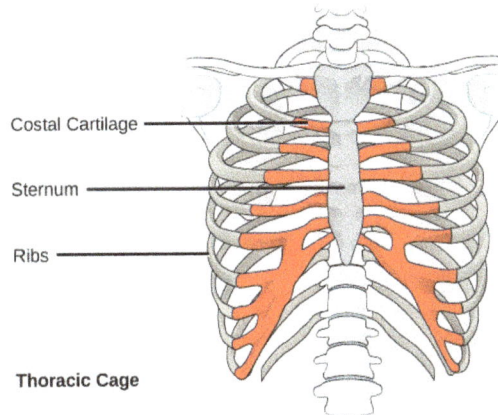

Figure: The thoracic cage, or rib cage, protects the heart and the lungs

The sternum, or breastbone, is a long, flat bone located at the anterior of the chest. It is formed from three bones that fuse in the adult. The ribs are 12 pairs of long, curved bones that attach to the thoracic vertebrae and curve toward the front of the body, forming the ribcage. Costal cartilages connect the anterior ends of the ribs to the sternum, with the exception of rib pairs 11 and 12, which are free-floating ribs.

Human Head

Head, in human anatomy, the upper portion of the body, consisting of the skull with its coverings and contents, including the lower jaw. It is attached to the spinal column by way of the first cervical vertebra, the atlas, and connected with the trunk of the body by the muscles, blood vessels, and nerves that constitute the neck. The term also is used to describe the anterior or fore part of animals other than humans.

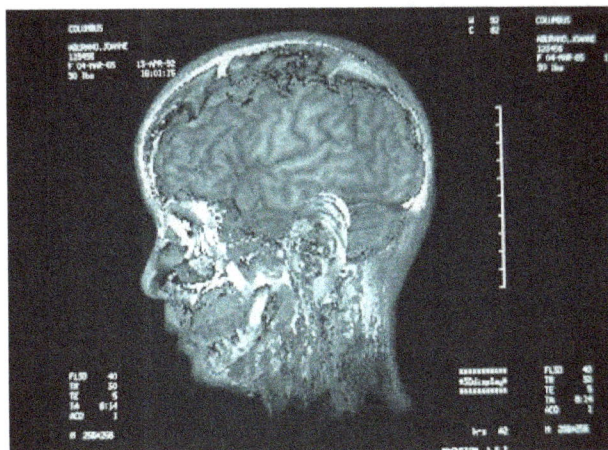

An MRI of a human head

The head is composed of the cranial and facial parts. It contains the brain, the centre of the nervous system. The cranium is partly covered with hair. The parts of the face are the forehead, the temples, the ears, the eyes with eyebrows, the cheeks, the nose, the jaw, the mouth and the chin. The main parts of the mouth are lips, the tongue, the teeth, the palate and the gums. The eyes are protected by eyelids and eyelashes. The eyes are the organs of sight. The nose of smell, and the ears are the organs of hearing. The nerves of the skin are organs of touch. The five senses are: sight, hearing, smell, taste and touch.

The head is attached to the trunk by the neck. The trunk includes the chest (in front), the back, the shoulders and the abdomen.

Structure

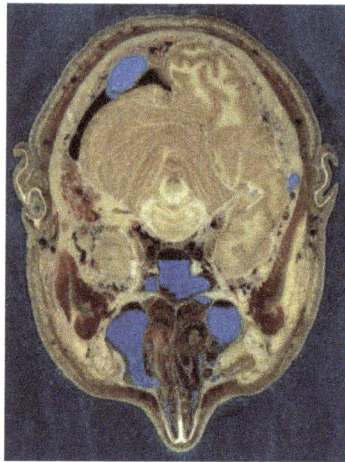

Cryosection through the male head

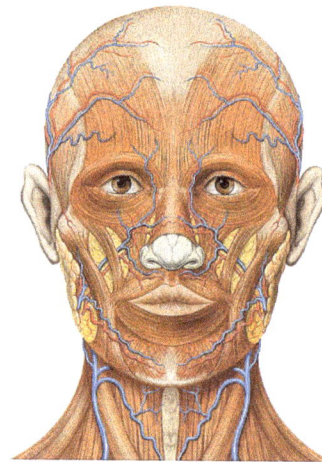

Anatomy of the human head

The human head consists of a fleshy outer portion surrounding the bony skull, within which sits the brain. The head rests on the neck, and is provided bony support for movement by the seven cervical vertebrae.

The face is the anterior part of the head, containing the sensory organs the eyes, nose and mouth. The cheeks, on either side of the mouth, provide a fleshy border to the oral cavity. To either side of the head sit the ears.

Blood Supply

The head receives blood supply through the internal and external carotid arteries. These supply the area outside the skull (external carotid artery) and inside of the skull (internal carotid artery). The area inside the skull also receives blood supply from the vertebral arteries, which travel up through the cervical vertebrae.

Nerve Supply

The twelve pairs of cranial nerves provide the majority of nervous control to the head. The sensation to the face is provided by the branches of the trigeminal nerve, a cranial nerve. The sensation to other portions of the head is provided by the cervical nerves.

Modern texts are in agreement about which areas of the skin are served by which nerves, but there are minor variations in some of the details. The borders designated by the diagrams in the 1918 edition of Gray's Anatomy, provided below, are similar but not identical to those generally accepted today.

The cutaneous innervation of the head is as follows:

- Ophthalmic nerve (green)

- Maxillary nerve (pink)

- Mandibular nerve (yellow)

- Cervical plexus (purple)

- Dorsal rami of cervical nerves (blue)

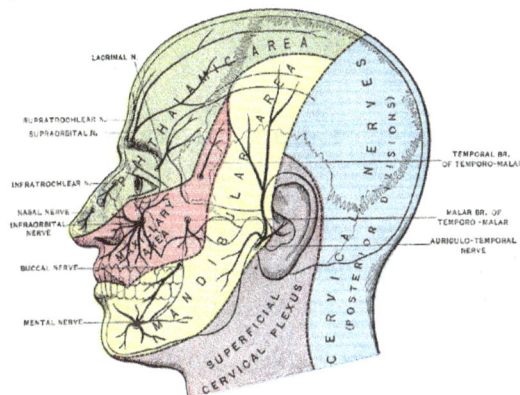

Sensory areas of the head, showing the general distribution of the three divisions of the fifth nerve

Function

The head contains sensory organs: two eyes, two ears, a nose and inside the mouth a tongue. It also houses the brain.

Headache

Headache is defined as a pain arising from the head or upper neck of the body. The pain originates from the tissues and structures that surround the skull or the brain because the brain itself has no nerves that give rise to the sensation of pain (pain fibers). The thin layer of tissue (periosteum) that surrounds bones, muscles that encase the skull, sinuses, eyes, and ears, as well as thin tissues that cover the surface of the brain and spinal cord (meninges), arteries, veins, and nerves, all can become inflamed or irritated and cause headache. The pain may be a dull ache, sharp, throbbing, constant, intermittent, mild, or intense.

Classification of Headaches

In 2013, the International Headache Society released its latest classification system for headache. Because so many people suffer from headaches, and because treatment is difficult sometimes, it was hoped that the new classification system would help health-care professionals make a more specific diagnosis as to the type of headache a patient has, and allow better and more effective options for treatment.

The guidelines are extensive and the Headache Society recommends that health-care professionals consult the guidelines frequently to make certain of the diagnosis.

There are three major categories of headache based upon the source of the pain.

1.   Primary headaches

2.   Secondary headaches

3.   Cranial neuralgias, facial pain, and other headaches

The guidelines also note that a patient may have symptoms that are consistent with more than one type of headache, and that more than one type of headache may be present at the same time.

Primary Headaches

Primary headaches include migraine, tension, and cluster headaches, as well as a variety of other less common types of headache.

- Tension headaches are the most common type of primary headache. Tension headaches occur more commonly among women than men. According to the World Health Organization, 1 in 20 people in the developed world suffer with a daily tension headache.

- Migraine headaches are the second most common type of primary headache. Migraine headaches affect children as well as adults. Before puberty, boys and girls are affected equally by migraine headaches, but after puberty, more women than men are affected.

- Cluster headaches are a rare type of primary headache. It more commonly affects men in their late 20s though women and children can also suffer from this type of headache.

Primary headaches can affect the quality of life. Some people have occasional headaches that resolve quickly while others are debilitating. While these headaches are not life threatening, they may be associated with symptoms that can mimic strokes.

Many patients equate severe headache with migraine, but the amount of pain does not determine the diagnosis of migraine.

Secondary Headaches

Secondary headaches are those that are due to an underlying structural or infectious problem in the head or neck. This is a very broad group of medical conditions ranging from dental pain from infected teeth or pain from an infected sinus, to life-threatening conditions like bleeding in the brain or infections like encephalitis or meningitis.

Traumatic headaches fall into this category including post-concussion headaches.

This group of headaches also includes those headaches associated with substance abuse and excess use of medications used to treat headaches (medication overuse headaches). "Hangover" headaches fall into this category as well. People who drink too much alcohol may waken with a well-established headache due to the effects of alcohol and dehydration.

Cranial Neuralgias, Facial Pain, and other Headaches

Neuralgia means nerve pain (neur=nerve + algia=pain). Cranial neuralgia describes inflammation of one of the 12 cranial nerves coming from the brain that control the muscles and carry sensory signals (such as pain) to and from the head and neck. Perhaps the most commonly recognized example is trigeminal neuralgia, which affects cranial nerve V (the trigeminal nerve), the sensory nerve that supplies the face and can cause intense facial pain when irritated or inflamed.

Appendicular Skeleton

The appendicular skeleton is composed of the bones of the upper limbs (which function to grasp and manipulate objects) and the lower limbs (which permit locomotion). It also includes the pectoral girdle, or shoulder girdle, that attaches the upper limbs to the body, and the pelvic girdle that attaches the lower limbs to the body (Figure below).

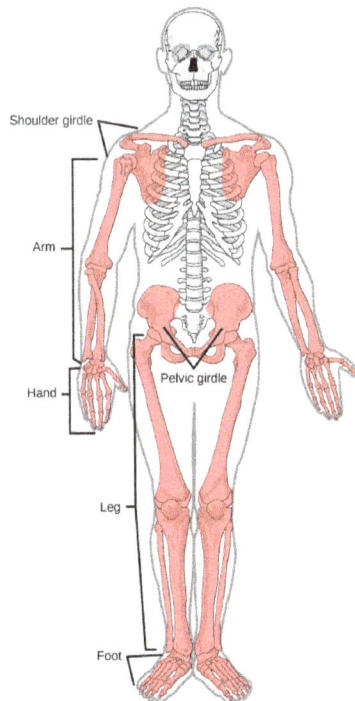

Figure: The appendicular skeleton is composed of the bones of the pectoral limbs (arm, forearm, hand), the pelvic limbs (thigh, leg, foot), the pectoral girdle, and the pelvic girdle

The Pectoral Girdle

The pectoral girdle bones provide the points of attachment of the upper limbs to the axial skeleton. The human pectoral girdle consists of the clavicle (or collarbone) in the anterior, and the scapula (or shoulder blades) in the posterior.

The clavicles are S-shaped bones that position the arms on the body. The clavicles lie horizontally across the front of the thorax (chest) just above the first rib. These bones are fairly fragile and are susceptible to fractures. For example, a fall with the arms outstretched causes the force to be transmitted to the clavicles, which can break if the force is excessive. The clavicle articulates with the sternum and the scapula.

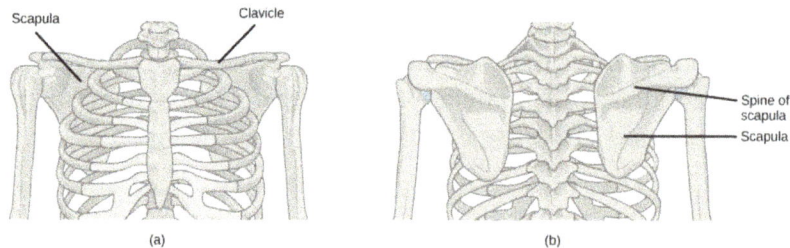

Figure: (a) The pectoral girdle in primates consists of the clavicles and scapulae. (b) The posterior view reveals the spine of the scapula to which muscle attaches.

The scapulae are flat, triangular bones that are located at the back of the pectoral girdle. They support the muscles crossing the shoulder joint. A ridge, called the spine, runs across the back of the scapula and can easily be felt through the skin (Figure above). The spine of the scapula is a good example of a bony protrusion that facilitates a broad area of attachment for muscles to bone.

The Upper Limb

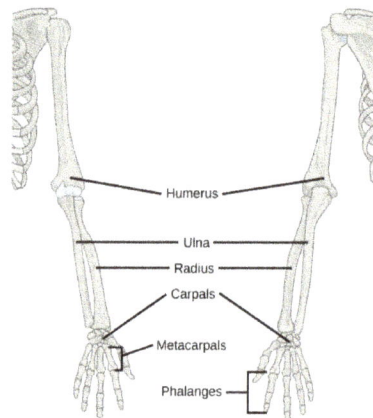

Figure: The upper limb consists of the humerus of the upper arm, the radius and ulna of the forearm, eight bones of the carpus, five bones of the metacarpus, and 14 bones of the phalanges

The upper limb contains 30 bones in three regions: the arm (shoulder to elbow), the forearm (ulna and radius), and the wrist and hand (Figure above).

An articulation is any place at which two bones are joined. The humerusis the largest and longest bone of the upper limb and the only bone of the arm. It articulates with the scapula at the shoulder and with the forearm at the elbow. The forearm extends from the elbow to the wrist and consists of two bones: the ulna and the radius. The radius is located along the lateral (thumb) side of the forearm and articulates with the humerus at the elbow. The ulna is located on the medial aspect (pinky-finger side) of the forearm. It is longer than the radius. The ulna articulates with the humerus at the elbow. The radius and ulna also articulate with the carpal bones and with each other, which in vertebrates enables a variable degree of rotation of the carpus with respect to the long axis of the limb. The hand includes the eight bones of the carpus (wrist), the five bones of the metacarpus(palm), and the 14 bones of the phalanges(digits). Each digit consists of three phalanges, except for the thumb, when present, which has only two.

The Pelvic Girdle

The pelvic girdle attaches to the lower limbs of the axial skeleton. Because it is responsible for bearing the weight of the body and for locomotion, the pelvic girdle is securely attached to the axial skeleton by strong ligaments. It also has deep sockets with robust ligaments to securely attach the femur to the body. The pelvic girdle is further strengthened by two large hip bones. In adults, the hip bones, or coxal bones, are formed by the fusion of three pairs of bones: the ilium, ischium, and pubis. The pelvis joins together in the anterior of the body at a joint called the pubic symphysis and with the bones of the sacrum at the posterior of the body.

The female pelvis is slightly different from the male pelvis. Over generations of evolution, females with a wider pubic angle and larger diameter pelvic canal reproduced more successfully. Therefore, their offspring also had pelvic anatomy that enabled successful childbirth (Figure below).

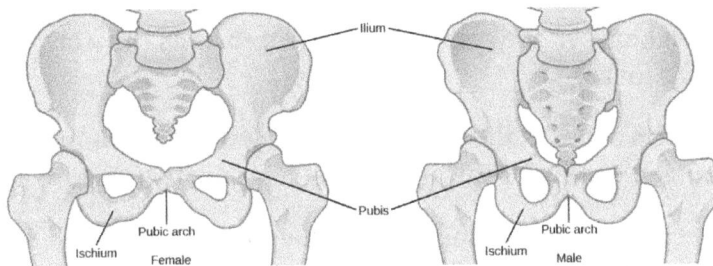

Figure: To adapt to reproductive fitness, the (a) female pelvis is lighter, wider, shallower, and has a broader angle between the pubic bones than (b) the male pelvis

The Lower Limb

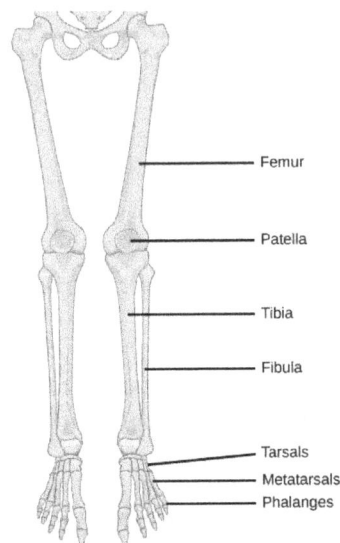

Figure: The lower limb consists of the thigh (femur), kneecap (patella), leg (tibia and fibula), ankle (tarsals), and foot (metatarsals and phalanges) bones

The lower limb consists of the thigh, the leg, and the foot. The bones of the lower limb are the femur (thigh bone), patella (kneecap), tibia and fibula (bones of the leg), tarsals (bones of the ankle), and metatarsals and phalanges (bones of the foot) (Figure above). The bones of the lower limbs are thicker and stronger than the bones of the upper limbs because of the need to support the entire weight of the body and the resulting forces from locomotion. In addition to evolutionary fitness, the bones of an individual will respond to forces exerted upon them.

The femur, or thighbone, is the longest, heaviest, and strongest bone in the body. The femur and pelvis form the hip joint at the proximal end. At the distal end, the femur, tibia, and patella form the knee joint. The patella, or kneecap, is a triangular bone that lies anterior to the knee joint. The patella is embedded in the tendon of the femoral extensors (quadriceps). It improves knee extension by reducing friction. The tibia, or shinbone, is a large bone of the leg that is located directly below the knee. The tibia articulates with the femur at its proximal end, with the fibula and the tarsal bones at its distal end. It is the second largest bone in the human body and is responsible for transmitting the weight of the body from the femur to the foot. The fibula, or calf bone, parallels and articulates with the tibia. It does not articulate with the femur and does not bear weight. The fibula acts as a site for muscle attachment and forms the lateral part of the ankle joint.

The tarsals are the seven bones of the ankle. The ankle transmits the weight of the body from the tibia and the fibula to the foot.

The metatarsals are the five bones of the foot. The phalanges are the 14 bones of the toes. Each toe consists of three phalanges, except for the big toe that has only two (Figure above).

Variations exist in other species; for example, the horse's metacarpals and metatarsals are oriented vertically and do not make contact with the substrate.

Shoulder Girdle

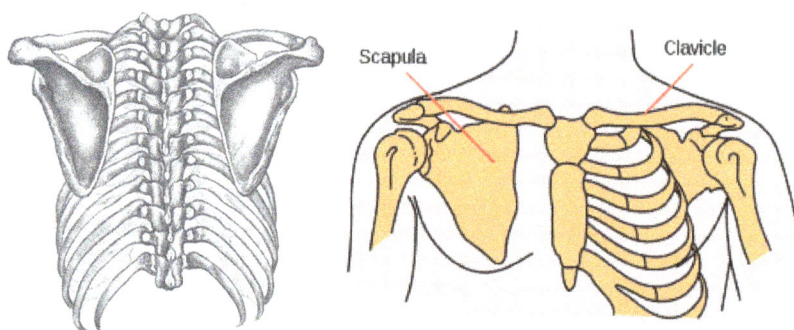

The shoulder girdle is four bones, two collar bones (clavicle) and two shoulder blades (scapula). These four bones make up the shoulder girdle as well as the shoulder socket. The shoulder girdle ideally hangs from the head suspended over the rib cage without either the collar bones or scapula touching the ribs.

The shoulder girdle has only one bony connection to the trunk at the front of the chest— the sterno clavicular joint where the collar bones meets the sternum. At the back of the body the shoulder blades are connected by the rhomboid muscles that stretch from the outer edge of the shoulder blade to the spine. As a result the shoulder girdle is two pair of bony units- right scapula and clavicle and left scapula and clavicle— as opposed to the seemingly solid pelvis.

The shoulder girdle is structurally similar to the pelvis but functionally different. The shoulder girdles mobility in the arms vs. the weight support function of the pelvis blurs their similarities. Because it is lighter and bound in only one place without firm support the shoulder girdle is more vulnerable to deformation then the pelvis. And being farther from the ground it offers gravity greater leverage for destruction.

The shoulder is made up of three bones—the collar bone, the shoulder blade, and the humerus, the arm bone. The direction and quality of arm movement is determined by muscles that form a bridge between shoulder and humerus. If pectoralis major is too tight the head of the humerus will be rotated in. If the latissimus dorsi is tight the humerus will be rotated internally. Teres major also has a role in determining the arms movement potential.

The shoulder girdle is designed to allow for very free movement of the arm and shoulder because in standing the upper arm is no longer involved in weight bearing. As a result its stability has been sacrificed for a greater range of motion. While the hips socket is a solid cup for the head of the femur to sit into, the shoulder socket is only a loose structure formed by the acromial and coracoid processes of the shoulder blade.

The arm and shoulder are so loose that a strange thing happens to allow for all of the available range of motion. Tendons of certain muscles take on the role of ligaments that would be too strong and tight to enable to ultimate freedom that the shoulder allows.

In Humans

The shoulder girdle is the anatomical mechanism that allows for all upper arm and shoulder movement in humans. The shoulder girdle consists of five muscles that attach to the clavicle and scapula and allow for the motion of the sternoclavicular joint (connection between sternum and clavicle) and acromioclavicular joint (connection between clavicle and scapula). The five muscles that comprise the function of the shoulder girdle are the trapezius muscle (upper, middle, and lower), levator scapulae muscle, rhomboid muscles (major and minor), serratus anterior muscle, and pectoralis minor muscle.

Joints

The shoulder girdle is a complex of five joints that can be divided into two groups. Three of these joints are true anatomicaljoints while two are physiological ("false") joints. Within each group, the joints are mechanically linked so that both groups simultaneously contribute to the different movements of the shoulder to variable degrees.

In the first group, the scapulohumeral or glenohumeral joint is the anatomical joint mechanically linked to the physiological subdeltoid or suprahumeral joint (the "second shoulder joint") so that movements in the suprahumeral joint results in movements in the glenohumeral joint. In the second group, the scapulocostal or scapulothoracic joint is the important physiological joint that can not function without the two anatomical joints in the group, the acromioclavicular and sternoclavicular joints, i.e. they join both ends of the clavicle.

Glenohumeral Joint

The glenohumeral joint is the articulation between the head of the humerus and the glenoid cavity of the scapula. It is a ball and socket type of synovial joint with three rotatory and three translatory degree of freedom. The glenohumeral joint allows for adduction, abduction, medial and lateral rotation, flexion and extension of the arm.

Acromioclavicular Joint

The acromioclavicular joint is the articulation between the acromion process of the scapula and the lateral end of the clavicle. It is a plane type of synovial joint. The acromion of the scapula rotates on the acromial end of the clavicle.

Sternoclavicular Joint

The sternoclavicular joint is the articulation of the manubrium of the sternum and the first costal cartilage with the medial end of the clavicle. It is a saddle type of synovial joint but functions as a plane joint. The sternoclavicular joint accommodates a wide range of scapula movements and can be raised to a 60° angle.

Scapulocostal Joint

The scapulocostal joint (also known as the scapulothoracic joint) is a physiological joint formed by an articulation of the anterior scapula and the posterior thoracic rib cage. It is musculotendinous in nature and is formed predominantly by the trapezius, rhomboids and serratus anterior muscles. The pectoralis minor also plays a role in its movements. The gliding movements at the scapulocostal joint are elevation, depression, retraction, protraction and superior and inferior rotation of the scapula. Disorders of the scapulocostal joint are not very common and usually restricted to snapping scapula.

Suprahumeral Joint

The suprahumeral joint (also known as the subacromial joint) is a physiological joint formed by an articulation of the coracoacromial ligament and the head of the humerus. It is formed by the gap between the humerus and the acromion process of the scapula. This space is filled mostly by the subacromial bursa and the tendon of supraspinatus. This joint plays a role during complex movements while the arm is fully flexed at the glenohumeral joint, such as changing a lightbulb, or painting a ceiling.

Movements

From its neutral position, the shoulder girdle can be rotated about an imaginary vertical axis at the medial end of the clavicle (the sternoclavicular joint). Throughout this movement the scapula is rotated around the chest wall so that it moves 15 centimetres (5.9 in) laterally and the glenoid cavity is rotated 40–45° in the horizontal plane. When the scapula is moved medially it lies in a frontal plane with the glenoid cavity facing directly laterally. At this position, the lateral end of the clavicle is rotated posteriorly so that the angle at the acromioclavicular joint opens up slightly. When the scapula is moved laterally it lies in a sagittal plane with the glenoid cavity facing anteriorly. At this position, the lateral end of the clavicle is rotated anteriorly so that the clavicle lies in a frontal plane. While this slightly closes the angle between the clavicle and the scapula, it also widens the shoulder.

The scapula can be elevated and depressed from the neutral position to a total range of 10 to 12 centimetres (3.9 to 4.7 in); at its most elevated position the scapula is always tilted so that the glenoid cavity is facing superiorly. During this tilting, the scapula rotates to a maximum angle of 60° about an axis passing perpendicularly through the bone slightly below the spine; this causes the inferior angle to move 10 to 12 centimetres (3.9 to 4.7 in) and the lateral angle 5 to 6 centimetres (2.0 to 2.4 in).

Injury

Shoulders are a common place for tissue injuries, especially if the person plays overhead sports such as tennis, volleyball, baseball, swimming, etc. According to Bahr's major injury related statistics, shoulder dislocations or subluxations account for 4% of injuries in adults ages 20–30 and 20% of shoulder injuries are fractures. Damage to the shoulder and adjacent features can fluctuate in severity depending on the person's age, sport, position, recurring shoulder dysfunction, and many other factors. Some other common shoulder injuries are fractures to any shoulder girdle bones i.e. clavicle, ligamentous sprains such as AC joint or GH ligaments, rotator cuff injuries, different labral tears, and other acute or chronic conditions related to shoulder anatomy.

Shoulder girdle pain can be acute or chronic and be due to a number of causes. Inflammation or injury of associated tendons, bone, muscles, nerves, ligaments, and cartilage can all cause pain. Also, past injury compensation, and stress can result in complicated shoulder pain.

Disorders

Winged Scapula

A Winged Scapula occurs for different reasons the two main reasons: palsy of the serratus anterior caused by a lesion on the Long Thoracic Nerve which is the more common or a lesion on the Spinal accessory nerve causing palsy in the trapezius muscle. These lesions can be caused by major trauma to the nerve, surgical procedure complication, as well as from under use of the serratus anterior or trapezius. The occurrence of this can be Unilateral or Bilateral both scapulae do not have to both be affected.

Serratus Anterior Muscle Palsy

As mentioned it is caused by a lesion on the long thoracic nerve this leads to a weakening of the serratus anterior on the medial border of the scapula. This separates Long thoracic nerve from spinal accessory nerve lesions.

Trapezius Muscle Palsy

Major Cause is a lesion on the Spinal Accessory Nerve this palsy presents differently than a lesion to the LTN. The issues caused in the trapezius show as a more mild detachment of the medial border of the scapula and a slide laterally away from the thoracic vertebrate.

Clavicle

Clavicle, also called Collarbone, curved anterior bone of the shoulder (pectoral) girdle in vertebrates; it functions as a strut to support the shoulder.

The clavicle is present in mammals with prehensile forelimbs and in bats, and it is absent in sea mammals and those adapted for running. The wishbone, or furcula, of birds is composed of the two fused clavicles; a crescent-shaped clavicle is present under the pectoral fin of some fish. In man the two clavicles, on either side of the anterior base of the neck, are horizontal, S-curved rods that articulate laterally with the outer end of the shoulder blade (the acromion) to help form the shoulder joint; they articulate medially with the breastbone(sternum). Strong ligaments hold the clavicle in place at either end; the shaft gives attachment to muscles of the shoulder girdle and neck. The clavicle may be congenitally reduced or absent; its robustness varies with degree of muscledevelopment.

Scapula

Scapula, also called shoulder blade, either of two large bones of the shouldergirdle in vertebrates. In humans they are triangular and lie on the upper back between the levels of the second and eighth ribs. A scapula's posterior surface is crossed obliquely by a prominent ridge, the spine, which divides the bone into two concave areas, the supraspinous and infraspinous fossae. The spine and fossae give attachment to muscles that act in rotating the arm. The spine ends in the acromion, a process that articulates with the clavicle, or collarbone, in front and helps form the upper part of the shoulder socket. The lateral apex of the triangle is broadened and presents a shallow cavity, the glenoid cavity, which articulates with the head of the bone of the upper arm, the humerus, to form the shoulder joint. Overhanging the glenoid cavity is a beaklike projection, the coracoid process, which completes the shoulder socket. To the margins of the scapula are attached muscles that aid in moving or fixing the shoulder as demanded by movements of the upper limb.

Arms and Forearms

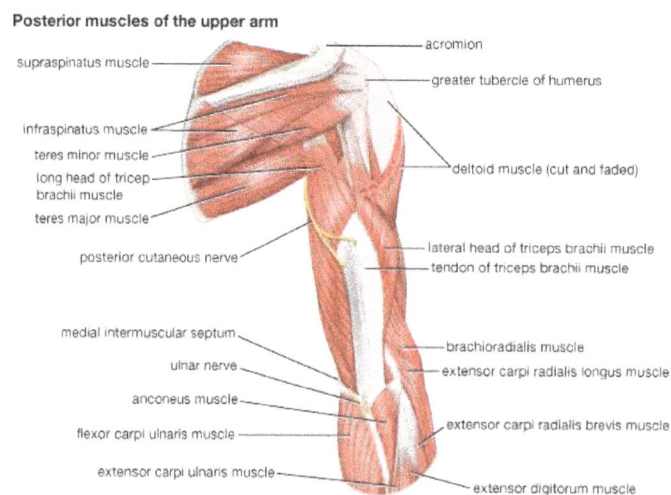

Posterior muscles of the upper arm

- acromion
- greater tubercle of humerus
- supraspinatus muscle
- infraspinatus muscle
- teres minor muscle
- long head of tricep brachii muscle
- teres major muscle
- deltoid muscle (cut and faded)
- posterior cutaneous nerve
- lateral head of triceps brachii muscle
- tendon of triceps brachii muscle
- medial intermuscular septum
- ulnar nerve
- anconeus muscle
- flexor carpi ulnaris muscle
- extensor carpi ulnaris muscle
- brachioradialis muscle
- extensor carpi radialis longus muscle
- extensor carpi radialis brevis muscle
- extensor digitorum muscle

Muscles of the upper arm (posterior view)

Arm, in zoology, either of the forelimbs or upper limbs of ordinarily bipedal vertebrates, particularly humans and other primates. The term is sometimes restricted to the proximal part, from shoulder to elbow (the distal part is then called the forearm). In brachiating (tree-swinging) primates the arm is unusually long.

The bones of the human arm, like those of other primates, consist of one long bone, the humerus, in the arm proper; two thinner bones, the radius and ulna, in the forearm; and sets of carpal and metacarpal bones in the hand and digits in the fingers. The muscle that extends, or straightens, the arm is the triceps, which arises on the humerus and attaches to the ulna at the elbow; the brachialis and biceps muscles act to bend the arm at the elbow. A number of smaller muscles cover the radius and ulna and act to move the hand and fingers in various ways. The pectoralis muscle, anchored in the chest, is important in the downward motion of the entire arm and in quadrupeds pulls the limb backward in locomotion.

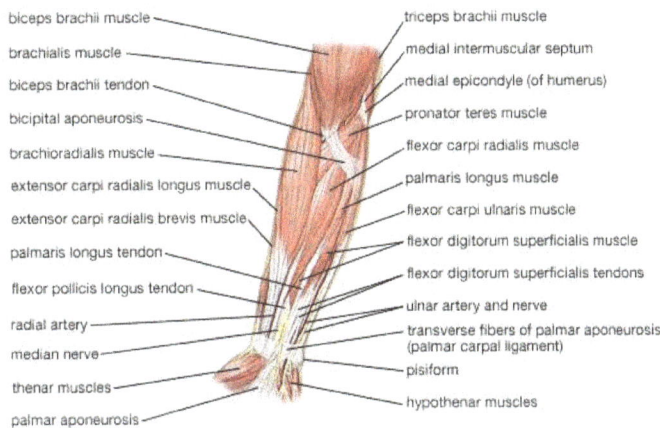

Muscles of the upper arm (posterior view)

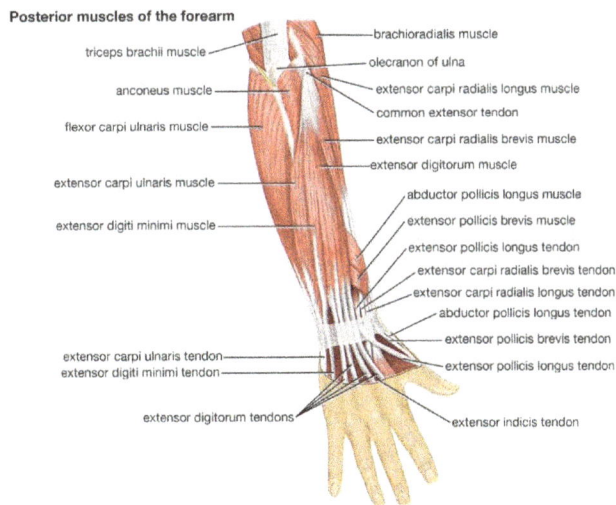

Muscles of the forearm (posterior view)

The term arm may also denote the limb or the locomotive or prehensile organ of an invertebrate, such as the ray of a starfish, tentacle of an octopus, or brachium of a brachiopod.

Overall, the forearm comprises the lower half of the arm. It extends from the elbow joint to the hand, and it is made up of the ulna and radius bones. These two long bones form a rotational joint, allowing the forearm to turn so that the palm of the hand faces up or down. The forearm is covered by skin, which provides a sensory function. The quantity of hair on the forearm varies for different people, but typically the top features more follicles than the underside. Two large arteries run the distance of the forearm, and these are the radial and ulnar. Both blood vessels follow a course close to the bones of similar name. These vessels also branch into lesser arteries, servicing the forearm's musculature. Many of the forearm's components are innervated by the radial, ulnar, and median nerves, as well as their distal branches. On the whole, the entire arm features three long bones that are frequently broken. This is especially true for the forearm. Often, a person extends their forearm instinctually when trying to break a fall, and this ultimately leads to an arm fracture.

Humerus

Humerus, long bone of the upper limb or forelimb of land vertebrates that forms the shoulder joint above, where it articulates with a lateral depression of the shoulder blade (glenoid cavity of scapula), and the elbow joint below, where it articulates with projections of the ulna and the radius.

In humans the articular surface of the head of the humerus is hemispherical; two rounded projections below and to one side receive, from the scapula, muscles that rotate the arm. The shaft is triangular in cross section and roughened where muscles attach. The lower end of the humerus includes two smooth articular surfaces (capitulum and trochlea), two depressions (fossae) that form part of the elbow joint, and two projections (epicondyles). The capitulum laterally articulates with the radius; the trochlea, a spool-shaped surface, articulates with the ulna. The two depressions—the olecranon fossa, behind and above the trochlea, and the coronoid fossa, in front and above—receive projections of the ulna as the elbow is alternately straightened and flexed. The epicondyles, one on either side of the bone, provide attachment for muscles concerned with movements of the forearm and fingers.

Ulna

The ulna is the longer, larger and more medial of the lower arm bones. Many muscles in the arm and forearm attach to the ulna to perform movements of the arm, hand and wrist. Movement of the ulna is essential to such everyday functions as throwing a ball and driving a car.

The ulna extends through the forearm from the elbow to the wrist, narrowing significantly towards its distal end. At its proximal end it forms the elbow joint with the humerus of the upper arm and the radius of the forearm. The ulna extends past the humerus to form the tip of the elbow, known as the olecranon.

The olecranon fits into a small recess in the humerus known as the olecranon fossa, preventing the elbow's extension beyond around 180 degrees. Just distal to the olecranon is the concave trochlear notch that surrounds the trochlea of the humerus to form the hinge of the elbow joint. The distal lip of the trochlear notch protrudes anteriorly to form the coronoid process that helps to lock the ulna in place with the humerus at the elbow and fits into the coronoid fossa of the humerus. On the lateral edge of the coronoid process is the small radial notch that forms the proximal radioulnar joint with the radius and permits the radius to rotate around the ulna at the elbow. A long ridge on the anterior side of the coronoid process known as the ulnar tuberosity extends down the shaft of the ulna as a muscle attachment point.

Moving distally from the elbow, the ulna begins to taper slightly in diameter along its entire length while curving medially. At its distal end, the ulna forms a small part of the wrist with the radius and the carpals of the hand. A rounded process known as the head of the ulna forms the distal radioulnar joint with the concave ulnar notch of the radius. The alignment of these joint-forming processes allows the radius to rotate around the ulna at the wrist. A small bony extension known as the styloid process protrudes from the posterior and medial corner of the ulna and provides an attachment point for the ulnar collateral ligament of the wrist.

Functionally, the ulna provides muscle attachment sites for over a dozen muscles in the upper arm and forearm. In the upper arm, the triceps brachii and anconeus muscles form insertions at the olecranon to extend the forearm at the elbow. The brachialis muscle has its insertion on the coronoid process to flex the arm at the elbow. Many muscles that act on the hand and wrist have their origins on the ulna, including the pronators, supinators, flexors and extensors.

Like its neighbors the humerus and radius, the ulna is classified as a long bone because of its long, narrow shape. All long bones have a similar structure, with a hollow shaft surrounded by compact bone and reinforced at the ends with spongy bone.

The hollow medullary cavity at the center of the ulna is filled with a soft, greasy substance known as yellow bone marrow. Yellow bone marrow contains many adipocytes and stores energy for the body in the form of triglycerides, or fats.

Surrounding the medullary cavity is the hard, dense compact bone made of mineral matrix and collagen fibers. The mix of collagen and minerals, including calcium, gives the ulna its great strength and flexibility.

The ends of the ulna are reinforced by spongy bone that increases the strength of the compact bone near the joints without significantly increasing the mass of the bone. Each region of spongy bone is made of many thin columns known as trabeculae that act like the steel girders of a bridge to withstand the stresses placed on the bone. Red bone marrow is found in the spaces between the trabeculae and contains many stem cells that produce the body's blood cells.

On the joint-forming ends of the ulna are thin layers of hyaline cartilage that cover the compact bone and protect it from the stresses at the joints. Hyaline is as smooth as ice to help bones glide past each other at the joints. It is also rubbery to absorb the shocks of impacts at the joints. The outer surface of the ulna (except at the joints) is covered in a thin fibrous layer known as the periosteum. Periosteum is made of a dense weave of collagen fibers that extend into the tendons and ligaments that attach the ulna to the muscles and bones of the arm.

The ulna begins at birth as a long bony shaft, known as the diaphysis, capped with hyaline cartilage at both ends. At around 4 years of age, the hyaline at the distal end by the wrist begins to ossify and forms a small bony cap known as the distal epiphysis. A thin layer of hyaline, known as the epiphyseal plate or growth plate, remains between the diaphysis and the newly formed epiphysis. The ulna grows lengthwise into the growth plate, which in turn grows to prevent the fusion of the diaphysis and epiphysis. At around age 10, the proximal tip of the olecranon begins to ossify and forms the proximal epiphysis. These three bones continue to grow and remain separated by the epiphyseal plates until the end of puberty and the beginning of adulthood, when they fuse together to form a single, unified ulna. The site of the epiphyseal plate becomes known as the metaphysis in the mature bone.

Radius (Bone)

The radius is the more lateral and slightly shorter of the two forearm bones. It is found on the thumb side of the forearm and rotates to allow the hand to pivot at the wrist. Several muscles of the arm and forearm have origins and insertions on the radius to provide motion to the upper limb. These movements are essential to many everyday tasks such as writing, drawing, and throwing a ball.

The radius is located on the lateral side of the forearm between the elbow and the wrist joints. It forms the elbow joint on its proximal end with the humerus of the upper arm and the ulna of the forearm.

Although the radius begins as the smaller of the two forearm bones at the elbow, it widens significantly as it extends along the forearm to become much wider than the ulna at the wrist. A short cylinder of smooth bone forms the head of the radius where it meets the capitulum of the humerus and the radial notch of the ulna at the elbow. The head of the radius allows the forearm to flex and pivot at the elbow joint. Just distal to the head, the radius narrows considerably to form the neck of the radius before expanding medially to form the radial tuberosity, a bony process that serves as the insertion of the biceps brachii.

Distal to the elbow, the body of the radius continues in a straight line along the lateral side of the forearm before suddenly widening just above the wrist joint. At its wide distal end, the radius terminates in three smooth, concave surfaces that form the wrist joint with the ulna and the carpals of the hand. Two of these concavities meet with the scaphoid and lunate bones of the carpals to form the radiocarpal portion of the wrist joint. On the medial side, the ulnar notch of the radius forms the distal radioulnar joint with the ulna, allowing the radius to rotate around the ulna to supinate and pronate the hand and wrist. The styloid process — a small, pointy extension of bone — protrudes from the lateral edge of the radius to anchor the radial collateral ligament of the wrist.

One of the most important functions of the radius is anchoring the muscles of the upper arm and the forearm. The biceps brachii muscle of the upper arm forms its insertion at the radial tuberosity to flex and supinate the forearm at the elbow. The supinator, pronator teres, and pronator quadratus muscles of the forearm also form insertions on the radius to supinate and pronate the hand and wrist by rotating the distal end of the radius around the ulna. Several muscles that move the hand and digits — including the flexor pollicis longus and flexor digitorum superficialis muscles — also have their origins on the radius.

The radius is classified structurally as a long bone because it is much longer than it is wide. Like all long bones, the radius is made of compact bone with a hollow center and spongy bones filling the ends. Compact bone is the hardest and heaviest part of the radius and makes up most of its structure. Many layers of minerals and collagen fibers give the compact bone its strength and flexibility.

Deep to the compact bone is a hollow cavity that spans the length of the bone and is filled with adipose-rich yellow bone marrow. Yellow bone marrow stores energy for the body's cells in the form of triglycerides.

At the proximal and distal end of the radius, the compact bone is reinforced by thin columns of spongy bone that give the radius extra strength without significantly adding to its mass. Small hollow spaces in the spongy bone house red bone marrow tissue that produces all of the body's blood cells.

The outer surface of the radius is covered in a thin layer of fibrous connective tissue known as the periosteum, and at its proximal and distal ends is covered in hyaline cartilage. Periosteum contains many collagen fibers that form strong connections between the radius and the tendons and ligaments that connect it to the bones and muscles of the arm. Hyaline cartilage gives the ends of the radius a smooth surface to reduce friction during movements of the forearm. It also acts as a flexible shock absorber to reduce impact stress at the elbow and wrist joints.

At birth, the radius begins as a bony shaft, known as the diaphysis, with a cap of hyaline cartilage on both ends. The hyaline cartilage provides extra flexibility to the elbow and wrist joints and provides a medium for the bone to grow into. Around the age of two, the distal hyaline cartilage near the wrist joint begins to turn into a separate bone called the distal epiphysis. A thin layer of cartilage called the epiphyseal plate (or growth plate) separates the diaphysis and epiphysis. The cartilage in the growth plate grows lengthwise to keep the diaphysis and epiphysis separated and to increase the overall length of the radius.

At around five years of age, the cartilage on the proximal end of the radius near the elbow ossifies to form the proximal epiphysis. Just like the distal epiphysis, an epiphyseal plate separates the proximal epiphysis from the diaphysis to give the radius room to grow. The epiphyses unite with the diaphysis by the end of puberty to form a single radius bone, at which point it stops growing lengthwise. The region where the diaphysis and epiphyses grow together is called the metaphysis.

Pelvis

Located in the lower torso, the pelvis is a sturdy ring of bones that protects the delicate organs of the abdominopelvic cavity while anchoring the powerful muscles of the hip, thigh, and abdomen. Several bones unite to form the pelvis, including the sacrum, coccyx (tail bone), and the left and right coxal (hip) bones.

Throughout childhood, the pelvis is made of many smaller bones that eventually fuse during adulthood to form a more rigid pelvis. Each of the coxal bones begins as three separate bones: the ilium, ischium, and pubis.

The ilium is the largest, widest, and most superior of the hip bones. When you place your hands on your hips, you can feel the curved ridge of the ilium known as the iliac crest. The narrow ischium is inferior to the ilium and is the bone, along with the coccyx, that you rest your body weight on while sitting. Anterior to the ischium is the pubis, the smallest of the hip bones. The ilium, ischium, and pubis meet in the center of the hip bone to form the deep, cup-like socket of the hip joint called the acetabulum.

The sacrum and coccyx also begin life as multiple bones before fusing. Five short, wide vertebrae fuse to form the wedge-shaped sacrum, while four tiny vertebrae fuse to form the coccyx.

The bones of the adult pelvis join together to form four joints: the left and right sacroiliac joints, the sacrococcygeal joint, and the pubic symphysis.

- The sacroiliac joints form between the sacrum and the left and right ilium to form a tight junction capable of supporting the body's weight and resisting the force of strong muscles. Although the sacroiliac joints are synovial joints, many strong ligaments and bony ridges on the sacrum and ilium interlock to prevent movement of the bones and strengthen the joint.

- The cartilaginous sacrococcygeal joint unites the sacrum with the tiny coccyx and allows some slight movement of the tail bone.

- On the anterior side of the pelvis the pubic symphysis unites the left and right pubic bones. The pubic symphysis is a slightly flexible band of fibrocartilage that allows the independent movement of the hip bones while walking. Women have a significantly wider and more flexible pubic symphysis compared to men, which allows the female pelvis to stretch during childbirth to accommodate the head of a fetus passing through the birth canal.

Many significant structural differences, or sexual dimorphisms, exist between the male and female pelvis. Most of these differences relate to the female reproductive system and the roles of pregnancy and childbirth in the female. The bones of the male pelvis are larger, thicker, and heavier than those of the female and there is very little hollow space within the ring of the pelvic bones. The female pelvis by comparison is significantly shorter and wider, which provides a greater hollow space within for the head of a fetus to pass through during childbirth. The sacrum and coccyx curve much more but are located more posteriorly in the female pelvis to increase the space of the pelvis. Even the shape and angle of the acetabulum and hip joint differ between males and females. The female acetabulum is rotated more anteriorly than in males, resulting in differences in walking gait and posture between men and women.

Hip Bone

Hip bone is also known as inominate bone or pelvic bone and is formed by fusion of three bones namely ilium, ischium and pubis bones. Hip bone forms part of pelvis and takes part in hip joint articulation.

The hip bone is made up of the three parts – the ilium, pubis and ischium. Prior to puberty, the triradiate cartilage separates these constituents. At the age of 15-17, the three parts begin to fuse.

Their fusion forms a cup-shaped socket known as the acetabulum, which becomes complete at 20-25 years of age. The head of the femur articulates with the acetabulum to form the hip joint.

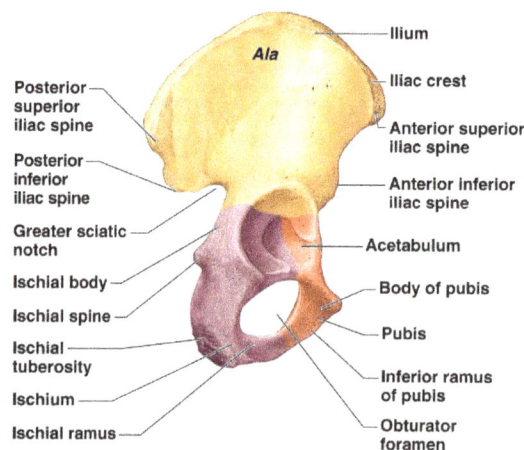

Thus the hip bone has three articulations –

- Sacroiliac joint – articulation with sacrum

- Pubic symphysis – articulation with the other hip bone.
- Hip joint – articulation with the head of femur.

Side Determination of Hip Bone

- Acetabulum is on lateral side.
- Obturator formen lies below the acetabulum, pubis being anterior and ischium posterior.
- Flat expanded part, the ilium is above the acetabulum.

Anatomical Position of Hip Bone

- Pubic tubercle and anterior superior iliac spine lie in same coronal plane.
- Pelvic surface of pubis is directed backwards and upwards.
- Symphyseal surface of the body of pubis is in the median plane.

Ilium

Ilium is largest part of the hip bone and forms upper expanded plate in upper part and contributes to acetabulum formation in lower part. Roughly two fifth of acetabulum is contributed by ilium.

Upper end of ilium is called iliac crest. Iliac crest is broad, convex, topmost portion of ilium which can be palpated in the flank area.

Anterior end of iliac crest is called the anterior superior iliac spine which is a very important anatomical landmark.

Iliac crest ends posteriorly in the posterior superior iliac spine [Located about dimple of venus about by a dimple 4 cm lateral to the second sacral spine.]

Iliac crest is divided in to a ventral segment and a dorsal segment which meet at the tubercle. The ventral segment forms more than the anterior two thirds of the crest. The dorsal segment forms less than the posterior one third of the crest.

Lower end is fused with ischium and pubis at the acetabulum.

Ilium has got three borders [anterior, posterior and medial] and three surfaces [gluteal, iliac, and a sacropelvic surface].

Borders of Ilium

Anterior border of ilium starts at the anterior superior iliac spine and runs downwards to the acetabulum. In upper part , the border has a notch and while its lower part has anterior inferior iliac spine.

Posterior border extends from the posterior superior iliac spine to the upper end of the posterior border of the ischium. It is marked by prominence called posterior inferior iliac spine and lower a large deep notch called the greater sciatic notch.

Medial Border is on the inner surface of the ilium from the iliac crest to the iliopubic eminence and separates the iliac fossa from the sacropelvic surface. Its lower rounder part form the iliac part inlet of pelvis or arcuate line.

Surfaces of Ilium

Gluteal surface is the outer surface of the ilium, which is convex in front and concave behind. Three gluteal lines divide the gluteal surface.

- The posterior gluteal line is the shortest and extends from a point front of the posterior superior spine to a point of the posterior inferior spine.

- The anterior gluteal line is the longest , begins about an inch behind the anterior superior spine, runs backwards and then downwards to end at the middle of the upper border of the greater sciatic notch.

- The inferior gluteal line is not well defined. It begins a little above and behind the anterior inferior spine, runs backwards and downwards to end near the apex of the greater sciatic notch.

Inner surface of ilium is divided by medial border into iliac fossa and sacropelvic surface.

Iliac fossa is the large concave area on the inner surface of the ilium, situated in front of its medial border. It forms the lateral wall of the false pelvis.

Sacropelvic surface is the uneven area on the inner surface of the ilium, behind its medial border. It is subdivided in three parts: the iliac tuberosity, auricular surface and pelvic surface.

Iliac tuberosity is a large rough area below dorsal segment of iliac crest. It is raised in the middle and depressed both above and below. Articular surface is called auricular surface and articulates with corresponding auricular surface on scarum to form sacroiliac joint.

Anteroinferior to auricular surface is smooth pelvic surface which forms lateral wall of true pelvis. Preauricular surface is marked by preauricular sulcus.

Attachments on Ilium

Attachments on external aspect of hip bone

Attachments on internal aspect of hip bone

Anterior Superior Iliac Spine

ASIS provides attachment to inguinal ligament. Sartorius muscle originates from ASIS and area below that.

Iliac Crest

Outer lip of iliac crest gives origin to fascia lata in whole extent and tensor fasica lata in front of the tubercle.

Latissimus dorsi muscle takes origin just behind the highest point of crest. Anterior two thirds of crest provides origin to external oblique muscles.

Inner lip of iliac crest gives rise to transverses abdominis muscle and fascia transversalis in anterior part and quadratus lumborum in posterior part.

In dorsal segment of iliac crest

- Gluteus maximus arises from lateral slope.
- Erector spinae arises from medial slope.
- Interosseus and dorsal sacroiliac ligaments are attached to medial margin deep to erector spinae.
- Anterior inferior iliac spine gives origin to straight head of rectus femoris and iliofemoral ligament.
- Sacrotuberous ligament gives origin to upper fibers of sacrotuberous ligament and few fibers of piriformis.

Gluteal Surface

- Gluteus maximus is a big muscle and its upper fibres take origin from area behind the posterior gluteal lines.
- From area between anterior and posterior gluteal lines, gluteus medius arises.
- Gluteus minimus arises from area between anterior and inferiorgluteal lines.
- Rectus femoris, reflected head arises from groove above the acetabulum.
- Capsule of the hip joint is attached to acetabular margin.

Iliac Fossa

- Iliac fossa gives origin to iliacus in upper two thirds where as lower one third is covered by iliac bursa.
- Iliac tuberosity provides attachments to interosseus sacroiliac ligament, dorsal sacroiliac ligament and iliolumbar ligament superiorly.
- Margin of auricular surface gives attachment to ventral sacroiliac ligament.

Pelvic Surface

Preauricular sulcus provides atttachement to lower fibers of ventralsacroiliac ligament, few fibers of piriformis and upper half of obturator internus.

Pubis

Pubis forms anteroinferior part of hip bone, contributes to anterior one fifth of acetabulum, and forms anterior boundary of obturator foramen.

Pubis consists of body, superior ramus and inferior ramus.

Body of Pubis

It is flattened anteroposteriorly and has a border superiorly called pubic crest which ends in a pubic tubercle laterally. In males the tubercle is crossed by the spermatic cord.

It has three surfaces

- Anterior surface is is directed downwards, forward and slightly laterally. It is rough on superomedial aspect. Posterior surface or pelvic surface is directed upwards and backwards and forms anterior wall of true pelvis.

- Medial surface or symphyseal surface articulates with opposite pubic symphysis.

Superior and inferior rami are extension from body of pubis.

Superior Ramus

Three borders

- Superior border [also called pectin pubis or pectineal line] is sharp and extends from pubic tubercle to posterior aspect of iliopubic eminence. It forms part of arcuate line. [pelvic inlet]

- The anterior border is called the obturator crest. This border is rounded ridge, extending form the public tubercle to the acetabular notch.

- The inferior border is sharp and forms the upper margin of the obturator foramen.

Three Surfaces

- The pectineal surface is a triangular area between the anterior and superior borders, extending from the pubic tubercle to the iliopubic eminence.

- The pelvic surface lies between the superior and inferior borders. It is smooth and is continuous with the pelvic surface of the body of the pubis. Pelvic surface is crossed by ductus deferens in males and round ligament of uterus in females.

- The obturator surface or inner surface lies between the anterior and inferior borders. It present the obturator groove. Obturator groove transmits obturator vessel and nerves.

Inferior Ramus

Inferior ramus extends from the body of the pubis to the ramus of the ischium, medial to the obturator foramen. It unites with the ramus of the ischium to form the conjoined ischiopubic rami.

Thighs and Legs

The thigh bears much of the load of the body's weight when a person is upright. It contains many muscles and nerves but only has one bone, the femur, which is the longest and strongest bone in the human body.

The four muscles that make up the quadriceps are the strongest and leanest of all muscles in the body. These muscles at the front of the thigh are the major extensors (help to extend the leg straight) of the knee. They are:

- Vastus lateralis
- Vastus medialis
- Vastus intermedius
- Rectus femoris

These four muscles come together to form a single tendon, which inserts into the patella, or kneecap.

Other muscles of the anterior (front) thigh include the pectineus, sartorius, and theiliopsoas, which is made up of the psoas major and iliacus.

Muscles in the medial thigh help to bring the thigh toward the midline of the body and rotate it. These muscles are the adductor longus, adductor brevis, adductor magnus, gracilis, and the obturator externus.

The hamstrings are three muscles at the back of the thigh that affect hip and knee movement. They begin under the gluteus maximus behind the hipbone and attach to the tibia at the knee. They are:

- Biceps femoris
- Semimembranosus
- Semitendinosus

Nerve supply to the thigh comes from various lumbar and sacral nerves via the femoral, obturator, and common peroneal nerves. The tibial and sciatic nerves also supply parts of the thigh.

The only bone in the thigh is the femur, which extends from the hip to the knee. It can resist forces of 1,800 to 2,500 pounds, so it is not easily fractured.

Branches of the femoral artery supply the thigh with oxygen-rich blood. The femoral artery is divided into a superficial, deep, and common arteries, and these further divide into branches, including the medial and lateral circumflex arteries. The largest branch of the femoral artery is the deep femoral artery, also called the profunda femoris. The femoral vein runs alongside the femoral artery and also has many branches. It takes oxygen-depleted blood from the thigh on a path back toward the heart.

Common problems with the thigh are often the result of participation in sports or repetitive movements. These include:

- Muscle strains (pulls or tears)

- Muscle cramps

- Contusions (bruises)

- Tendonitis (inflammation of a tendon)

- Sciatica (pain from the sciatic nerve)

Thigh pain is a common injury, but that does not mean it can't be serious. While sometimes pain and discomfort is unavoidable, arming yourself with accurate information can protect you. The best way to manage chronic thigh pain is to learn about the location, causes, symptoms, and treatment options.

Understanding Thigh Pain

Pain in the thigh may be caused by conditions that affect the ligaments, tendons, muscles, joints, bones, nerves, blood vessels, and skin. When left untreated, thigh pain can lead to potentially life-threatening complications.

Location of Thigh Pain

The precise location of thigh pain can vary depending on the underlying issue. Once your doctor has determined the reason behind your painful thighs, your treatment plan will focus on pain relief and controlling the root cause.

- Front Thigh Pain

Pain in front of the thigh is known as anterior thigh pain. Upper front thigh pain can happen suddenly and may be caused by muscle strains or contusions from a direct blow. Chronic or gradual onset of front thigh pain may occur if an injury has not been treated correctly.

- Back Thigh Pain

Pain in the back of the thigh is called posterior thigh pain. It can be sudden and acute, or it may be chronic and develop slowly. Back thigh pain may also occur after an injury that fails to heal properly.

- Outer Thigh Pain

The cause of pain in the outer thigh is sometimes obvious, such as a pulled muscle during a vigorous workout. However, outer thigh pain can also be due to less obvious conditions, such as a pinched nerve.

- Inner Thigh Pain

Inner thigh pain can be different for each person. How pain in the upper thigh presents itself depends on the root cause, but most people define the pain as a kind of jolt that keeps them awake at night. An obvious cause of inner and upper thigh pain is a pulled inner thigh muscle, but other causes are not related to physical activity.

Thigh Pain Causes

What causes thigh pain? There are different reasons a person may experience mid-thigh pain or lower thigh pain. Some appear suddenly after a specific incident, while others develop gradually. Here are common causes of thigh pain:

- Quadriceps Strain: The muscles in our thighs are made up of three major groups: the adductors, the hamstrings, and the quadriceps. A tear in the quadriceps is the most common cause of sudden pain in the front of the thigh. Quadriceps strains typically develop during kicking, jumping, or sprinting.

- Hamstring Contusion: Upper thigh pain causes discomfort, especially when the area is touched. If there is stiffness, bruising, and swelling, the pain may be caused by a hamstring contusion. A bruise develops when the muscle is crushed against the thigh bone. Hamstring contusions can range from mild to severe.

- Bursitis: Inflammation or irritation of the bursa typically causes intense pain in the upper, outer thigh. For those with bursitis in the knee, certain activities, such as standing from a seated position or climbing stairs, can be painful.

- Referred Pain: If your knee and thigh pain is associated with hip, back, or glute pain, referred pain is often the cause. The term describes pain that is felt in one location though it originates from a problem elsewhere.

- Meralgia Paresthetica: Whether it is left thigh pain or right thigh pain, meralgia paresthetica occurs when too much pressure is put on a nerve in the pelvic area. When this nerve is pinched, the feeling in the upper thigh is affected, resulting in thigh pain. Meralgia paresthetica can be easily confused with other conditions.

- Avulsion Fracture: An avulsion fracture is one of the less common causes of thigh pain. It happens because of excessive tension where a tendon or ligament attaches to the bone, which results in a bony fragment. An avulsion fracture is associated with sharp pain, loss of function, and swelling.

Symptoms of Thigh Muscle Pain

The symptoms of thigh muscle pain are often worse with prolonged standing and walking or during activities that require repetitive hip extension. The sensations vary, and discomfort is usually alleviated by lying down or sitting. Symptoms associated with thigh pain include:

- Severe pain in calf and thigh when walking or going up and down stairs

- Shooting pain in thigh and knee following a high-impact collision

- Numbness and burning pain in back of thigh

- Bruising, swelling, or tenderness

- Weakness and popping sensation at the time of injury

Diagnosing Sharp Pain in Thigh

Pain in thigh bones can be characterized as stabbing, with the severity and duration depending on the cause. A dull, aching pain in the thigh rarely requires a doctor visit. Most cases are simple muscular injuries that heal with rest and at-home treatments.

However, there are certain conditions and circumstances that require medical advice. To diagnose thigh pain, your doctor will conduct a thorough physical examination to look for signs of a serious condition. If further assessment is required, an MRI scan or ultrasound may be used to confirm the severity and location of your injury.

Leg, limb or appendage of an animal, used to support the body, provide locomotion, and, in modified form, assist in capturing and eating prey (as in certain shellfish, spiders, and insects). In four-limbed vertebrates all four appendages are commonly called legs, but in bipedal animals, including humans, only the posterior or lower two are so called.

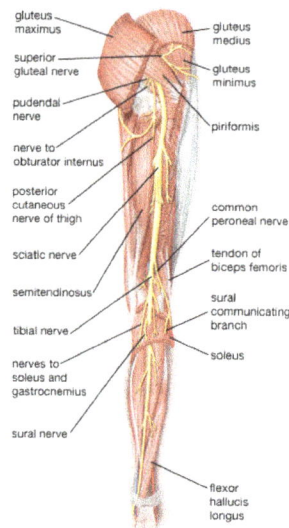

Posterior view of the right leg, showing the sciatic nerve and its branches

The bones of the human leg, like those of other mammals, consist of a basal segment, the femur (thighbone); an intermediate segment, the tibia (shinbone) and the smaller fibula; and a distal segment, the pes (foot), consisting of tarsals, metatarsals, and phalanges (toes).

In birds and bats the foreleg has evolved into the wing. Various other adaptations of the leg include modifications for swimming, digging, leaping, and running, as seen in the porpoise, the mole, the kangaroo, and the horse, respectively. The appendages of many invertebrates are also known as legs.

Structure

In human anatomy, the lower leg is the part of the lower limb that lies between the knee and the ankle. The thigh is between the hip and knee and makes up the rest of the lower limb. The term lower limb or "lower extremity" is commonly used to describe all of the leg.

The leg from the knee to the ankle is called the *crus* or *cnemis* /ˈniːmɪs/. The calf is the back portion, and the tibia or shinbone together with the smaller fibula make up the front of the lower leg.

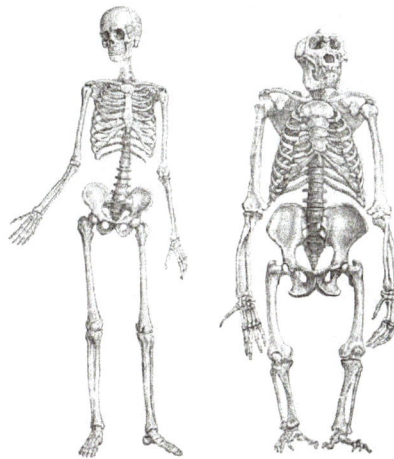

Comparison between human and gorilla skeletons. (Gorilla in non-natural stretched posture.)

Evolution has provided the human body with two distinct features: the specialization of the upper limb for visually guided manipulation and the lower limb's development into a mechanism specifically adapted for efficient bipedal gait. While the capacity to walk upright is not unique to humans, other primates can only achieve this for short periods and at a great expenditure of energy. The human adaption to bipedalism is not limited to the leg, however, but has also affected the location of the body's center of gravity, the reorganisation of internal organs, and the form and biomechanism of the trunk. In humans, the double S-shaped vertebral column acts as a shock-absorber which shifts the weight from the trunk over the load-bearing surface of the feet. The human legs are exceptionally long and powerful as a result of their exclusive specialization to support and locomotion — in orangutans the leg length is 111% of the trunk; in chimpanzees 128%, and in humans 171%. Many of the leg's muscles are also adapted to bipedalism, most substantially the gluteal muscles, the extensors of the knee joint, and the calf muscles.

Skeleton

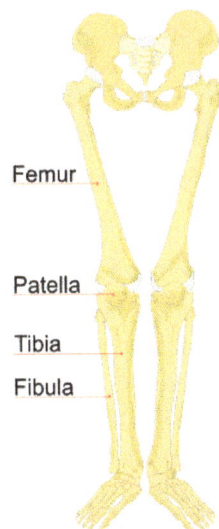

Femur

Patella

Tibia

Fibula

Bones of the leg

The major bones of the leg are the femur (thigh bone), tibia (shin bone), and adjacent fibula, and these are all long bones. The patella (kneecap) is the sesamoid bone in front of the knee. Most of the leg skeleton has bony prominences and margins that can be palpated and some serve as anatomical landmarks that define the extent of the leg. These landmarks are the anterior superior iliac spine, the greater trochanter, the superior margin of the medial condyle of tibia, and the medial malleolus. Notable exceptions to palpation are the hip joint, and the neck and body, or shaft of the femur.

Usually, the large joints of the lower limb are aligned in a straight line, which represents the mechanical longitudinal axis of the leg, the Mikulicz line. This line stretches from the hip joint(or more precisely the head of the femur), through the knee joint (the intercondylar eminenceof the tibia), and down to the center of the ankle (the ankle mortise, the fork-like grip between the medial and lateral malleoli). In the tibial shaft, the mechanical and anatomical axes coincide, but in the femoral shaft they diverge 6°, resulting in the *femorotibial angle* of 174° in a leg with normal axial alignment. A leg is considered straight when, with the feet brought together, both the medial malleoli of the ankle and the medial condyles of the knee are touching. Divergence from the normal femorotibial angle is called genu varum if the center of the knee joint is lateral to the mechanical axis (intermalleolar distance exceeds 3 cm), and genu valgum if it is medial to the mechanical axis (intercondylar distance exceeds 5 cm). These conditions impose unbalanced loads on the joints and stretching of either the thigh's adductors and abductors. The angle of inclination formed between the neck and shaft of the femur, (collodiaphysial angle), varies with age—about 150° in the newborn, it gradually decreases to 126-128° in adults, to reach 120° in old age. Pathological changes in this angle results in abnormal posture of the leg: A small angle produces coxa vara and a large angle in coxa valga; the latter is usually combined with genu varum and coxa vara leads genu valgum. Additionally, a line drawn through the femoral neck superimposed on a line drawn through the femoral condyles forms an angle, the *torsion* angle, which makes it possible for flexion movements of the hip joint to be transposed into rotary movements of the femoral head. Abnormally increased torsion angles results in a limb turned inward and a decreased angle in a limb turned outward; both cases resulting in a reduced range of a persons mobility.

Muscles

Hip

There are several ways of classifying the muscles of the hip: (1) By location or innervation (ventral and dorsal divisions of the plexus layer); (2) by development on the basis of their points of insertion (a posterior group in two layers and an anterior group); and (3) by function (i.e. extensors, flexors, adductors, and abductors).

Some hip muscles also act on either the knee joint or on vertebral joints. Additionally, because the area of origin and insertion of many of these muscles are very extensive, these muscles are often involved in several very different movements. In the hip joint, lateral and medial rotation occur along the axis of the limb; extension (also called dorsiflexion or retroversion) and flexion (anteflexion or anteversion) occur along a transverse axis; and abduction and adduction occur about a sagittal axis.

The anterior dorsal hip muscles are the iliopsoas, a group of two or three muscles with a shared insertion on the lesser trochanter of the femur. The psoas major originates from the last vertebra and along the lumbar spine to stretch down into the pelvis. The iliacus originates on the iliac fossa on the interior side of the pelvis. The two muscles unite to form the iliopsoas muscle which is inserted on the lesser trochanter of the femur. The psoas minor, only present in about 50 per cent of subjects, originates above psoas major to stretch obliquely down to its insertion on the interior side of the major muscle.

The posterior dorsal hip muscles are inserted on or directly below the greater trochanter of the femur. The tensor fasciae latae, stretching from the anterior superior iliac spine down into the iliotibial tract, presses the head of the femur into the acetabulum but also flexes, rotates medially, and abducts to hip joint. The piriformis originates on the anterior pelvic surface of the sacrum, passes through the greater sciatic foramen, and inserts on the posterior aspect of the tip of the greater trochanter. In a standing posture it is a lateral rotator, but it also assists extending the thigh. The gluteus maximus has its origin between (and around) the iliac crest and the coccyxfrom where one part radiates into the iliotibial tract and the other stretches down to the gluteal tuberosity under the greater trochanter. The gluteus maximus is primarily an extensor and lateral rotator of the hip joint, and it comes into action when climbing stairs or rising from a sitting to standing posture. Furthermore, the part inserted into the fascia latae abducts and the part inserted into the gluteal tuberosity adducts the hip. The two deep glutei muscles, the gluteus medius and minimus, originate on the lateral side of the pelvis. The medius muscle is shaped like a cap. Its anterior fibers act as a medial rotator and flexor; the posterior fibers as a lateral rotator and extensor; and the entire muscle abducts the hip. The minimus has similar functions and both muscles are inserted onto the greater trochanter.

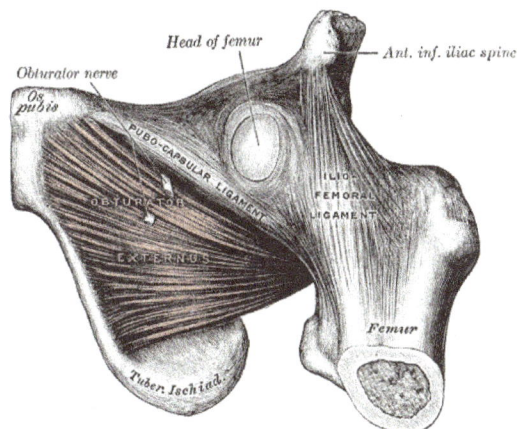

Muscles of hip

The ventral hip muscles function as lateral rotators and play an important role in the control of the body's balance. Because they are stronger than the medial rotators, in the normal position of the leg, the apex of the foot is pointing outward to achieve better support. The obturator internus originates on the pelvis on the obturator foramen and its membrane, passes through the lesser sciatic foramen, and is inserted on the trochanteric fossa of the femur. "Bent" over the lesser sciatic notch, which acts as a fulcrum, the muscle forms the strongest lateral rotators of the hip together with the gluteus maximus and quadratus femoris. When sitting with the knees flexed it acts as an abductor. The obturator externus has a parallel course with its origin located on the posterior border of the obturator foramen. It is covered by several muscles and acts as a lateral rotator and a weak adductor. The inferior and superior gemelli represent marginal heads of the obturator internus and assist this muscle. The three muscles have been referred to as the triceps coxae. The quadratus femoris originates at the ischial tuberosity and is inserted onto the intertrochanteric crest between the trochanters. This flattened muscle act as a strong lateral rotator and adductor of the thigh.

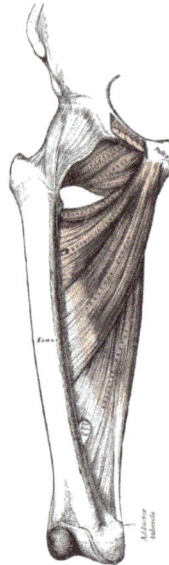

Hip adductors

The adductor muscles of the thigh are innervated by the obturator nerve, with the exception of pectineuswhich receives fibers from the femoral nerve, and the adductor magnus which receives fibers from the tibial nerve. The gracilis arises from near the pubic symphysis and is unique among the adductors in that it reaches past the knee to attach on the medial side of the shaft of the tibia, thus acting on two joints. It share its distal insertion with the sartorius and semitendinosus, all three muscles forming the pes anserinus. It is the most medial muscle of the adductors, and with the thigh abducted its origin can be clearly seen arching under the skin. With the knee extended, it adducts the thigh and flexes the hip. The pectineus has its origin on the iliopubic eminence laterally to the gracilis and, rectangular in shape, extends obliquely to attach immediately behind the lesser trochanter and down the pectineal line and the proximal part of the linea aspera on the femur. It is a flexor of the hip joint, and an adductor and a weak medial rotator of the thigh. The adductor brevis originates on the inferior ramus of the pubis below the gracilis and stretches obliquely below the pectineus down to the upper third of the linea aspera. Except for being an adductor, it is a lateral rotator and weak flexor of the hip joint. The adductor longus has its origin at superior ramus of the pubis and inserts medially on the middle third of the linea aspera. Primarily an adductor, it is also responsible for some flexion. The adductor magnus has its origin just behind the longus and lies deep to it. Its wide belly divides into two parts: One is inserted into the linea aspera and the tendon of the other reaches down to adductor tubercle on the medial side of the femur's distal end where it forms an intermuscular septum that separates the flexors from the extensors. Magnus is a powerful adductor, especially active when crossing legs. Its superior part is a lateral rotator but the inferior part acts as a medial rotator on the flexed leg when rotated outward and also extends the hip joint. The adductor minimus is an incompletely separated subdivision of the adductor magnus. Its origin forms an anterior part of the magnus and distally it is inserted on the linea aspera above the magnus. It acts to adduct and lateral rotate the femur.

Thigh

The muscles of the thigh can be classified into three groups according to their location: anterior and posterior muscles and the adductors (on the medial side). All the adductors except gracilis insert on the femur and act on the hip joint, and so functionally qualify as hip muscles. The majority of the thigh muscles, the "true" thigh muscles, insert on the leg (either the tibia or the fibula) and act primarily on the knee joint. Generally, the extensors lie on anterior of the thigh and flexors lie on the posterior. Even though the sartorius flexes the knee, it is ontogenetically considered an extensor since its displacement is secondary.

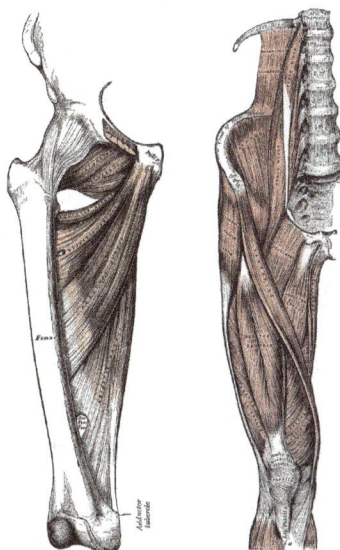

Anterior and posterior thigh muscles

Of the anterior thigh muscles the largest are the four muscles of the quadriceps femoris: the central rectus femoris, which is surrounded by the three vasti, the vastus intermedius, medialis, and lateralis. Rectus femoris is attached to the pelvis with two tendons, while the vasti are inserted to the femur. All four muscles unite in a common tendon inserted into the patella from where the patellar ligament extends it down to the tibial tuberosity. Fibers from the medial and lateral vasti form two retinacula that stretch past the patella on either sides down to the condyles of the tibia. The quadriceps is *the* knee extensor, but the rectus femoris additionally flexes the hip joint, and articular muscle of the knee protects the articular capsule of the knee joint from being nipped during extension. The sartorius runs superficially and obliquely down on the anterior side of the thigh, from the anterior superior iliac spine to the pes anserinus on the medial side of the knee, from where it is further extended into the crural fascia. The sartorius acts as a flexor on both the hip and knee, but, due to its oblique course, also contributes to medial rotation of the leg as one of the pes anserinus muscles (with the knee flexed), and to lateral rotation of the hip joint.

There are four posterior thigh muscles. The biceps femoris has two heads: The long head has its origin on the ischial tuberosity together with the semitendinosus and acts on two joints. The short head originates from the middle third of the linea aspera on the shaft of the femur and the lateral intermuscular septum of thigh, and acts on only one joint. These two heads unite to form the biceps which inserts on the head of the fibula. The biceps flexes the knee joint and rotates the flexed leg laterally — it is the only lateral rotator of the knee and thus has to oppose all medial rotator.

Additionally, the long head extends the hip joint. The semitendinosus and the semimembranosus share their origin with the long head of the biceps, and both attaches on the medial side of the proximal head of the tibia together with the gracilis and sartorius to form the pes anserinus. The semitendinosus acts on two joints; extension of the hip, flexion of the knee, and medial rotation of the leg. Distally, the semimembranosus' tendon is divided into three parts referred to as the *pes anserinus profondus*. Functionally, the semimembranosus is similar to the semitendinosus, and thus produces extension at the hip joint and flexion and medial rotation at the knee. Posteriorly below the knee joint, the popliteus stretches obliquely from the lateral femoral epicondyledown to the posterior surface of the tibia. The subpopliteal bursa is located deep to the muscle. Popliteus flexes the knee joint and medially rotates the leg.

Crus and Foot

With the popliteus as the single exception, all muscles in the leg are attached to the foot and, based on location, can be classified into an anterior and a posterior group separated from each other by the tibia, the fibula, and the interosseous membrane. In turn, these two groups can be subdivided into subgroups or layers — the anterior group consists of the extensors and the peroneals, and the posterior group of a superficial and a deep layer. Functionally, the muscles of the leg are either extensors, responsible for the dorsiflexion of the foot, or flexors, responsible for the plantar flexion. These muscles can also classified by innervation, muscles supplied by the anterior subdivision of the plexus and those supplied by the posterior subdivision. The leg muscles acting on the foot are called the extrinsic foot muscles whilst the foot muscles located *in* the foot are called intrinsic.

Dorsiflexion (extension) and plantar flexion occur around the transverse axis running through the ankle joint from the tip of the medial malleolus to the tip of the lateral malleolus. Pronation (eversion) and supination (inversion) occur along the oblique axis of the ankle joint.

Extrinsic

Anterior muscles

Three of the anterior muscles are extensors. From its origin on the lateral surface of the tibia and the interosseus membrane, the three-sided belly of the tibialis anterior extends down below the superior and inferior extensor retinacula to its insertion on the plantar side of the medial cuneiform bone and the first metatarsal bone. In the non-weight-bearing leg, the anterior tibialis dorsal flexes the foot and lifts the medial edge of the foot. In the weight-bearing leg, it pulls the leg towards the foot. The extensor digitorum longus has a wide origin stretching from the lateral condyle of the tibia down along the anterior side of the fibula, and the interosseus membrane. At the ankle, the tendon divides into four that stretch across the foot to the dorsal aponeuroses of the last phalanges of the four lateral toes. In the non-weight-bearing leg, the muscle extends the digits and dorsiflexes the foot, and in the weight-bearing leg acts similar to the tibialis anterior. The extensor hallucis longus has its origin on the fibula and the interosseus membrane between the two other extensors and is, similarly to the extensor digitorum, is inserted on the last phalanx of big toe ("hallux"). The muscle dorsiflexes the hallux, and acts similar to the tibialis anterior in the weight-bearing leg. Two muscles on the lateral side of the leg form the peroneal group. The peroneus longus and brevis both have their origins on the fibula and they both pass behind the lateral malleolus where their tendons pass under the peroneal retinacula. Under the foot, the longus stretches from the lateral to the medial side in a groove, thus bracing the transverse arch of the foot. The brevis is attached on the lateral side to the tuberosity of the fifth metatarsal. Together the two peroneals form the strongest pronators of the foot. The peroneus muscles are highly variable and several variants can occasionally be present.

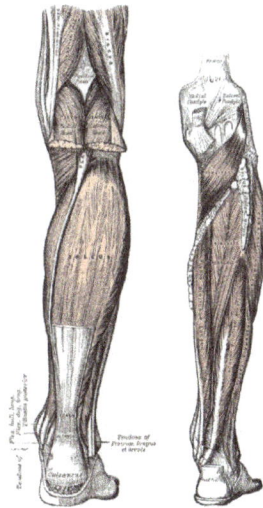

Superficial and deep posterior muscles.

Of the posterior muscles three are in the superficial layer. The major plantar flexors, commonly referred to as the triceps surae, are the soleus, which arises on the proximal side of both leg bones, and the gastrocnemius, the two heads of which arises on the distal end of the femur. These muscles unite in a large terminal tendon, the Achilles tendon, which is attached to the posterior tubercle of the calcaneus. The plantaris closely follows the lateral head of the gastrocnemius. Its tendon runs between those of the soleus and gastrocnemius and is embedded in the medial end of the calcaneus tendon.

In the deep layer, the tibialis posterior has its origin on the interosseus membrane and the neighbouring bone areas and runs down behind the medial malleolus. Under the foot it splits into a thick medial part attached to the navicular bone and a slightly weaker lateral part inserted to the three cuneiform bones. The muscle produces simultaneous plantar flexion and supination in the non-weight-bearing leg, and approximates the heel to the calf of the leg. The flexor hallucis longus arises distally on the fibula and on the interosseus membrane from where its relatively thick muscle belly extends far distally. Its tendon extends beneath the flexor retinaculum to the sole of the foot and finally attaches on the base of the last phalanx of the hallux. It plantarflexes the hallux and assists in supination. The flexor digitorum longus, finally, has its origin on the upper part of the tibia. Its tendon runs to the sole of the foot where it forks into four terminal tendon attached to the last phalanges of the four lateral toes. It crosses the tendon of the tibialis posterior distally on the tibia, and the tendon of the flexor hallucis longus in the sole. Distally to its division, the quadratus plantae radiates into it and near the middle phalanges its tendons penetrate the tendons of the flexor digitorum brevis. In the non-weight-bearing leg, it plantar flexes the toes and foot and supinates. In the weight-bearing leg it supports the plantar arch.

Intrinsic

The intrinsic muscles of the foot, muscles whose bellies are located in the foot proper, are either dorsal (top) or plantar (sole). On the dorsal side, two long extrinsic extensor muscles are superficial to the intrinsic muscles, and their tendons form the dorsal aponeurosis of the toes. The short intrinsic extensors and the plantar and dorsal interossei radiates into these aponeuroses. The extensor digitorum brevis and extensor hallucis brevis have a common origin on the anterior side of the calcaneus, from where their tendons extend into the dorsal aponeuroses of digits 1-4. They act to dorsiflex these digits.

The plantar muscles can be subdivided into three groups associated with three regions: those of the big digit, the little digit, and the region between these two. All these muscles are covered by the thick and dense plantar aponeurosis, which, together with two tough septa, form the spaces of the three groups. These muscles and their fatty tissue function as cushions that transmit the weight of the body downward. As a whole, the foot is a functional entity.

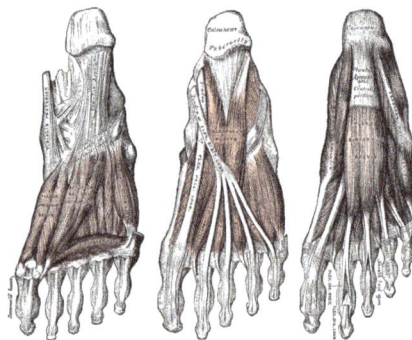

Intrinsic foot muscles

The abductor hallucis stretches along the medial edge of the foot, from the calcaneus to the base of the first phalanx of the first digit and the medial sesamoid bone. It is an abductor and a weak flexor, and also helps maintain the arch of the foot. Lateral to the abductor hallucis is the flexor hallucis brevis, which originates from the medial cuneiform bone and from the tendon of the tibialis posterior. The flexor hallucis has a medial and a lateral head inserted laterally to the abductor hallucis. It is an important plantar flexor which comes into prominent use in classical ballet (i.e. for pointe work). The adductor hallucis has two heads; a stronger oblique head which arises from the cuboid and lateral cuneiform bones and the bases of the second and third metatarsals; and a transverse head which arises from the distal ends of the third-fifth metatarsals. Both heads are inserted on the lateral sesamoid bone of the first digit. The muscle acts as a tensor to the arches of the foot, but can also adduct the first digit and plantar flex its first phalanx.

The opponens digiti minimi originates from the long plantar ligament and the plantar tendinous sheath of peroneus longus and is inserted on the fifth metatarsal. When present, it acts to plantar flex the fifth digit and supports the plantar arch. The flexor digiti minimi arises from the region of base of the fifth metatarsal and is inserted onto the base of the first phalanx of the fifth digit where it is usually merged with the abductor of the first digit. It acts to plantar flex the last digit. The largest and longest muscles of the little toe is the abductor digiti minimi. Stretching from the lateral process of the calcaneus, with a second attachment on the base of the fifth metatarsal, to the base of the fifth digit's first phalanx, the muscle forms the lateral edge of the sole. Except for supporting the arch, it plantar flexes the little toe and also acts as an abductor.

The four lumbricales have their origin on the tendons of the flexor digitorum longus, from where they extend to the medial side of the bases of the first phalanx of digits two-five. Except for reinforcing the plantar arch, they contribute to plantar flexion and move the four digits toward the big toe. They are, in contrast to the lumbricales of the hand, rather variable, sometimes absent and sometimes more than four are present. The quadratus plantae arises with two slips from margins of the plantar surface of the calcaneus and is inserted into the tendon(s) of the flexor digitorum longus, and is known as the "plantar head" of this latter muscle. The three plantar interossei arise with their single heads on the medial side of the third-fifth metatarsals and are inserted on the bases of the first phalanges of these digits. The two heads of the four dorsal interossei arise on two adjacent metatarsals and merge in the intermediary spaces. Their distal attachment is on the bases of the proximal phalanges of the second-fourth digits. The interossei are organized with the second digit as a longitudinal axis; the plantars act as adductors and pull digits 3-5 towards the second digit; while the dorsals act as abductors. Additionally, the interossei act as plantar flexors at the metatarsophalangeal joints. Lastly, the flexor digitorum brevis arises from underneath the calcaneus to insert its tendons on the middle phalanges of digit 2-4. Because the tendons of the flexor digitorum longus run between these tendons, the brevis is sometimes called *perforatus*. The tendons of these two muscles are surrounded by a tendinous sheath. The brevis acts to plantar flex the middle phalanges.

Flexibility

Flexibility can be simply defined as the available range of motion (ROM) provided by a specific joint or group of joints. For the most part, exercises that increase flexibility are performed with intentions to boost overall muscle length, reduce the risks of injury and to potentially improve muscular performance in physical activity. Stretching muscles after engagement in any physical activity can improve muscular strength, increase flexibility, and reduce muscle soreness. If limited movement is present within a joint, the "insufficient extensibility" of the muscle, or muscle group, could be restricting the activity of the affected joint.

Stretching

Stretching prior to strenuous physical activity has been thought to increase muscular performance by extending the soft tissue past its attainable length in order to increase range of motion. Many physically active individuals practice these techniques as a "warm-up" in order to achieve a certain level of muscular preparation for specific exercise movements. When stretching, muscles should feel somewhat uncomfortable but not physically agonizing.

- Plantar flexion: One of the most popular lower leg muscle stretches is the step standing heel raises, which mainly involves the gastrocnemius, soleus, and the Achilles tendon. Standing heel raises allow the individual to activate their calf muscles by standing on a step with toes and forefoot, leaving the heel hanging off the step, and plantar flexing the ankle joint by raising the heel. This exercise is easily modified by holding on to a nearby rail for balance and is generally repeated 5-10 times.

- Dorsiflexion: In order to stretch the anterior muscles of the lower leg, crossover shin stretches work well. This motion will stretch the dorsiflexion muscles, mainly the anterior tibialis, extensor hallucis longus and extensor digitorum longus, by slowly causing the muscles to lengthen as body weight is leaned on the ankle joint by using the floor as resistance against the top of the foot. Crossover shin stretches can vary in intensity depending on the amount of body weight applied on the ankle joint as the individual bends at the knee. This stretch is typically held for 15–30 seconds.

- Eversion and inversion: Stretching the eversion and inversion muscles allows for better range of motion to the ankle joint. Seated ankle elevations and depressions will stretch the peroneus and tibilalis muscles that are associated with these movements as they lengthen. Eversion muscles are stretched when the ankle becomes depressed from the starting position. In like manner, the inversion muscles are stretched when the ankle joint becomes elevated. Throughout this seated stretch, the ankle joint is to remain supported while depressed and elevated with the ipsilateral (same side) hand in order to sustain the stretch for 10–15 seconds. This stretch will increase overall eversion and inversion muscle group length and provide more flexibility to the ankle joint for larger range of motion during activity.

Blood Supply

The arteries of the leg are divided into a series of segments.

In the pelvis area, at the level of the last lumbar vertebra, the abdominal aorta, a continuation the descending aorta, splits into a pair of common iliac arteries. These immediately split into the internal and external iliac arteries, the latter of which descends along the medial border of the psoas major to exits the pelvis area through the vascular lacuna under the inguinal ligament.

The artery enters the thigh as the femoral artery which descends the medial side of the thigh to the adductor canal. The canal passes from the anterior to the posterior side of the limb where the artery leaves through the adductor hiatus and becomes the popliteal artery. On the back of the knee the popliteal artery runs through the popliteal fossa to the popliteal muscle where it divides into anterior and posterior tibial arteries.

In the lower leg, the anterior tibial enters the extensor compartment near the upper border of the interosseus membrane to descend between the tibialis anterior and the extensor hallucis longus. Distal to the superior and extensor retinacula of the foot it becomes the dorsal artery of the foot. The posterior tibial forms a direct continuation of the popliteal artery which enters the flexor compartment of the lower leg to descend behind the medial malleolus where it divides into the medial and lateral plantar arteries, of which the posterior branch gives rise to the fibular artery.

For practical reasons the lower limb is subdivided into somewhat arbitrary regions: The regions of the hip are all located in the thigh: anteriorly, the subinguinal region is bounded by the inguinal ligament, the sartorius, and the pectineus and forms part of the femoral triangle which extends distally to the adductor longus. Posteriorly, the gluteal region corresponds to the gluteus maximus. The anterior region of the thigh extends distally from the femoral triangle to the region of the knee and laterally to the tensor fasciae latae. The posterior region ends distally before the popliteal fossa. The anterior and posterior regions of the knee extend from the proximal regions down to the level of the tuberosity of the tibia. In the lower leg the anterior and posterior regions extend down to the malleoli. Behind the malleoli are the lateral and medial retromalleolar regions and behind these is the region of the heel. Finally, the foot is subdivided into a dorsal region superiorly and a plantar region inferiorly.

Veins

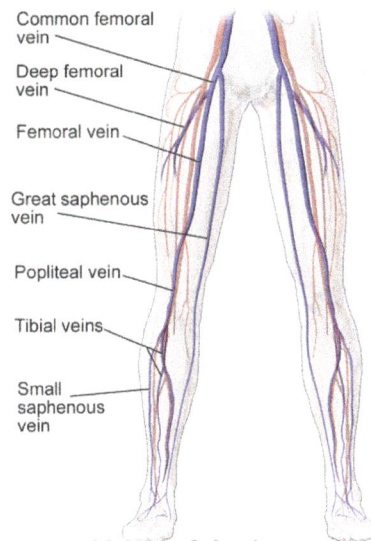

Common femoral vein
Deep femoral vein
Femoral vein
Great saphenous vein
Popliteal vein
Tibial veins
Small saphenous vein

Veins of the leg

The veins are subdivided into three systems. The deep veins return approximately 85 percent of the blood and the superficial veins approximately 15 percent. A series of perforator veinsinterconnect the superficial and deep systems. In the standing posture, the veins of the leg have to handle an exceptional load as they act against gravity when they return the blood to the heart. The venous valves assist in maintaining the superficial to deep direction of the blood flow.

- Superficial Veins:

Greater saphenous vein

Small saphenous

- Deep veins:

- Femoral vein

- Popliteal vein

- Anterior tibial vein

- Posterior tibial vein

- Fibular vein

Innervation

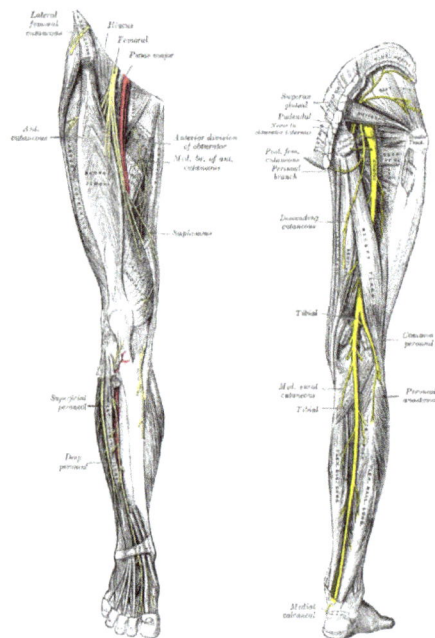

Nerves of right leg, anterior and posterior aspects

The sensory and motor innervation to the lower limb is supplied by the lumbosacral plexus, which is formed by the ventral rami of the lumbar and sacral spinal nerves with additional contributions from the subcostal nerve (T12) and coccygeal nerve (Co1). Based on distribution and topography, the lumbosacral plexus is subdivided into the lumbar plexus (T12-L4) and the Sacral plexus (L5-S4); the latter is often further subdivided into the sciatic and pudendal plexuses.

The lumbar plexus is formed lateral to the intervertebral foramina by the ventral rami of the first four lumbar spinal nerves (L1-L4), which all pass through psoas major. The larger branches of the plexus exit the muscle to pass sharply downward to reach the abdominal wall and the thigh (under the inguinal ligament); with the exception of the obturator nerve which pass through the lesser pelvis to reach the medial part of the thigh through the obturator foramen. The nerves of the lumbar plexus pass in front of the hip joint and mainly support the anterior part of the thigh.

The iliohypogastric (T12-L1) and ilioinguinal nerves (L1) emerge from the psoas major near the muscle's origin, from where they run laterally downward to pass anteriorly above the iliac crest between the transversus abdominis and abdominal internal oblique, and then run above the inguinal ligament. Both nerves give off muscular branches to both these muscles. Iliohypogastric supplies sensory branches to the skin of the lateral hip region, and its terminal branch finally pierces the aponeurosis of the abdominal external oblique above the inguinal ring to supply sensory branches to the skin there. Ilioinguinalis exits through the inguinal ring and supplies sensory branches to the skin above the pubic symphysis and the lateral portion of the scrotum.

The genitofemoral nerve (L1, L2) leaves psoas major below the two former nerves, immediately divides into two branches that descends along the muscle's anterior side. The sensory femoral branch supplies the skin below the inguinal ligament, while the mixed genital branch supplies the skin and muscles around the sex organ. The lateral femoral cutaneous nerve (L2, L3) leaves psoas major laterally below the previous nerve, runs obliquely and laterally downward above the iliacus, exits the pelvic area near the iliac spine, and supplies the skin of the anterior thigh.

The obturator nerve (L2-L4) passes medially behind psoas major to exit the pelvis through the obturator canal, after which it gives off branches to obturator externus and divides into two branches passing behind and in front of adductor brevis to supply motor innervation to all the other adductor muscles. The anterior branch also supplies sensory nerves to the skin on a small area on the distal medial aspect of the thigh. The femoral nerve (L2-L4) is the largest and longest of the nerves of the lumbar plexus. It supplies motor innervation to iliopsoas, pectineus, sartorius, and quadriceps; and sensory branches to the anterior thigh, medial lower leg, and posterior foot.

The nerves of the sacral plexus pass behind the hip joint to innervate the posterior part of the thigh, most of the lower leg, and the foot. The superior (L4-S1) and inferior gluteal nerves (L5-S2) innervate the gluteus muscles and the tensor fasciae latae. The posterior femoral cutaneous nerve (S1-S3) contributes sensory branches to the skin on the posterior thigh. The sciatic nerve (L4-S3), the largest and longest nerve in the human body, leaves the pelvis through the greater sciatic foramen. In the posterior thigh it first gives off branches to the short head of the biceps femoris and then divides into the tibial (L4-S3) and common fibular nerves (L4-S2). The fibular nerve continues down on the medial side of biceps femoris, winds around the fibular neck and enters the front of the lower leg. There it divides into a deep and a superficial terminal branch. The superficial branch supplies the peroneus muscles and the deep branch enters the extensor compartment; both branches reaches into the dorsal foot. In the thigh, the tibial nerve gives off branches to semitendinosus, semimembranosus, adductor magnus, and the long head of the biceps femoris. The nerve then runs straight down the back of the leg, through the popliteal fossa to supply the ankle flexors on the back of the lower leg and then continues down to supply all the muscles in the sole of the foot. The pudendal (S2-S4) and coccygeal nerves (S5-Co2) supply the muscles of the pelvic floor and the surrounding skin.

The lumbosacral trunk is a communicating branch passing between the sacral and lumbar plexuses containing ventral fibers from L4. The coccygeal nerve, the last spinal nerve, emerges from the sacral hiatus, unites with the ventral rami of the two last sacral nerves, and forms the coccygeal plexus.

Lower Leg and Foot

The lower leg and ankle need to keep exercised and moving well as they are the base of the whole body. The lower extremities must be strong in order to balance the weight of the rest of the body, and the gastrocnemius muscles take part in much of the blood circulation.

Exercises

Isometric and Standard

There are a number of exercises that can be done to strengthen the lower leg. For example, in order to activate plantar flexorsin the deep plantar flexors one can sit on the floor with the hips flexed, the ankle neutral with knees fully extended as they alternate pushing their foot against a wall or platform. This kind of exercise is beneficial as it hardly causes any fatigue.Another form of isometric exercise for the gastrocnemius would be seated calf raises which can be done with or without equipment. One can be seated at a table with their feet flat on the ground, and then plantar flex both ankles so that the heels are raised off the floor and the gastrocnemius flexed. An alternate movement could be heel drop exercises with the toes being propped on an elevated surface—as an opposing movement this would improve the range of motion. One-legged toe raises for the gastrocnemius muscle can be performed by holding one dumbbell in one hand while using the other for balance, and then standing with one foot on a plate. The next step would be to plantar flex and keep the knee joint straight or flexed slightly. The triceps surae is contracted during this exercise. Stabilization exercises like the BOSU ball squat are also important especially as they assist in the ankles having to adjust to the ball's form in order to balance.

Feet and Ankles

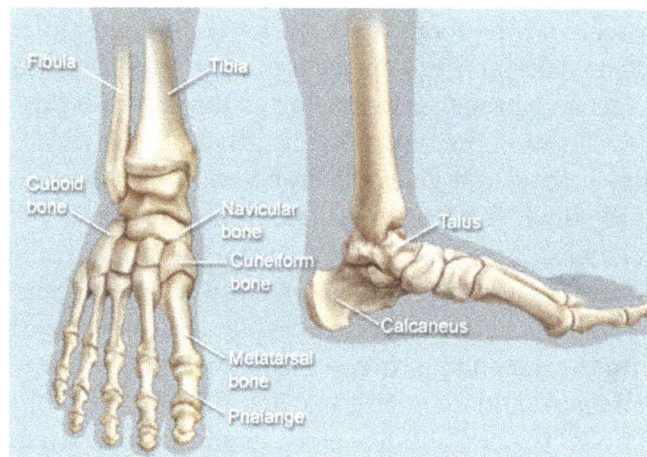

The feet are flexible structures of bones, joints, muscles, and soft tissues that let us stand upright and perform activities like walking, running, and jumping. The feet are divided into three sections:

- The forefoot contains the five toes (phalanges) and the five longer bones (metatarsals).

- The midfoot is a pyramid-like collection of bones that form the arches of the feet. These include the three cuneiform bones, the cuboid bone, and the navicular bone.

- The hindfoot forms the heel and ankle. The talus bone supports the leg bones (tibia and fibula), forming the ankle. The calcaneus (heel bone) is the largest bone in the foot.

Muscles, tendons, and ligaments run along the surfaces of the feet, allowing the complex movements needed for motion and balance. The Achilles tendon connects the heel to the calf muscle and is essential for running, jumping, and standing on the toes.

The ankle is a large joint made up of three bones:

- The shin bone (tibia)
- The thinner bone running next to the shin bone (fibula)
- A foot bone that sits above the heel bone (talus)

The bony bumps (or protrusions) seen and felt on the ankle have their own names:

- The medial malleolus, felt on the inside of your ankle is part of the tibia's base
- The posterior malleolus, felt on the back of your ankle is also part of the tibia's base
- The lateral malleolus, felt on the outside of your ankle is the low end of the fibula

The ankle joint allows up-and-down movement of the foot. The subtalar joint sits below the ankle joint, and allows side-to-side motion of the foot. Numerous ligaments (made of tough, moveable tissue) surround the true ankle and subtalar joints, binding the bones of the leg to each other and to those of the foot.

Tarsus (Skeleton)

Tarsal, any of several short, angular bones that in humans make up the ankle and that—in animals that walk on their toes (*e.g.,* dogs, cats) or on hoofs—are contained in the hock, lifted off the ground. The tarsals correspond to the carpal bones of the upper limb. In humans the tarsals, in combination with the metatarsal bones, form a longitudinal arch in the foot—a shape well adapted for carrying and transferring weight in bipedal locomotion.

In the human ankle there are seven tarsal bones. The talus (astragalus) articulates above with the bones of the lower leg to form the ankle joint. The other six tarsals, tightly bound together by ligaments below the talus, function as a strong weight-bearing platform. The calcaneus, or heel bone, is the largest tarsal and forms the prominence at the back of the foot. The remaining tarsals include the navicular, cuboid, and three cuneiforms. The cuboid and cuneiforms adjoin the metatarsal bones in a firm, nearly immovable joint.

Metatarsal Bones

Metatarsals are part of the bones of the mid-foot and are tubular in shape. They are named by numbers and start from the medial side outward. The medial side is the same side as the big toe.

They are called the first metatarsal, second metatarsal, third metatarsal, fourth metatarsal, and the fifth metatarsal. The first metatarsal is the strongest of the group.

These bones are found between the phalanges of the toes and the tarsal bones. Each bone's base will move with at least one of the tarsal bones where the tarsometatarsal joint is located. The metatarsal bones are connected to the bones of the toe, or phalanges, at the knuckle of the toe, or metatarsophalangeal joint.

Metatarsals are convex in shape (arch upward), are long bones, and give the foot its arch. They work with connective tissues, ligaments and tendons, to provide movement in the foot.

These bones can become fractured, strained, or inflamed from misuse or overuse. Immobilization of the foot (e.g. via casting) can help heal metatarsal fractures and sprains.

Phalanx Bone

Phalanx (plural: *phalanges*) refers to the bones found in fingers, toes, paws, wings, hooves and fins of animals. These are long bones whose length exceeds their breadth. Phalanges are connected to each other at hinge-like inter phalangeal joints that can be used for either flexion or extension. Some phalanges are fused to each other.

There are 56 phalanges in the human body, with the thumbs and large toes having 2 bones each. The remaining fingers and toes have 3 bones with the fourth and fifth toes having fused phalanges. This gives rise to the phalangeal formula of 2+3+3+3+3 bones per limb. Depending on their position, these bones are attached to the bones in the palm (metacarpals) and foot(metatarsals) through ligaments at the metacarpophalangeal and metatarsophalangeal joints. Tendons connect phalanges to muscles in the palm and forearm, which direct the movement of fingers. Similar attachments are seen for phalanges in the toes.

The word phalanx derives from the Greek term for a military formation where soldiers stood in several rows and columns. Each digit has two or more bones and every palm or foot has multiple digits. When these bone structures are seen (as in an X-ray or a skeleton), it can appear as if the bones are arranged side-by-side, two or three rows deep, like an army formation.

Types of Phalanges

Phalanges can be classified based on their position relative to the rest of the body. The bone closest to the metacarpals and metatarsals is a proximal phalanx, while the one farthest away (usually positioned below the nail) is a distal phalanx. Some digits have more than two bones and those situated between the proximal and distal bones are called intermediate phalanges. While human fingers and toes have either 2 or 3 phalanx bones, there are primitive reptiles with 4 and some marine mammals can have up to 12 bones in a digit.

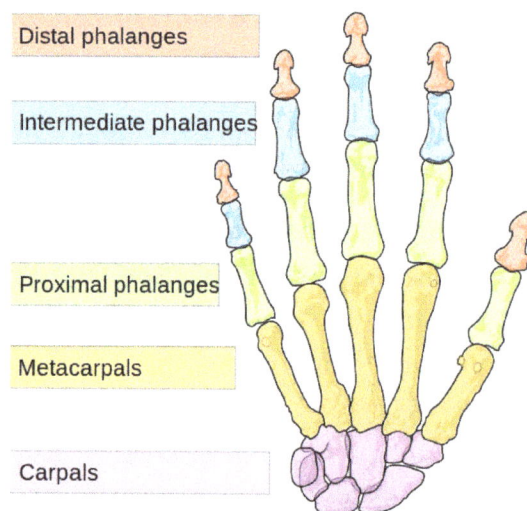

Distal phalanges

Intermediate phalanges

Proximal phalanges

Metacarpals

Carpals

Human hand bones

Type I: Proximal Phalanges

These bones form the base of fingers and toes and connect them to the rest of the limb. These phalanges also form the knuckles of the hand. In mammals like bats, the proximal phalanx connects to the metacarpophalangeal joint to form the base of the wing. Depending on the function, these bones can vary in shape. In humans, the proximal phalanx in the hand is relatively broad with a concave surface near the palm. The corresponding bone in the foot appears shorter, with a convex surface above.

In horses, the proximal phalanx is called the pastern and appears like an hourglass. It is nearly twice as long as the next intermediate phalanx. The position of the hoof during running, which depends on the nature of the surface, is supposed to affect the occurrence of fractures in these bones in racehorses. Proximal phalanges are also important in the study of fossils to understand locomotion patterns since they are the skeletal elements that are in close apposition with the substrate.

Injuries to proximal phalanges include fractures and dislocations. They are usually treated with realignment under local anesthetic and a splint to immobilize the bone and joint.

Type II: Intermediate Phalanges

Intermediate phalanges are not present in every digit, and are missing from the thumb and large toe in humans. These bones are intermediate in position, length and size. They are a part of two inter phalangeal joints, interacting with the proximal and distal phalanx bones. In horses, intermediate or middle phalanx bones are called coronets and form the upper limit of the hoof. The shape and flexibility of this phalanx is important for animals like bats, where the bone has to bend and flex with the flow of air, rather than restricting its movement.

Type III: Distal Phalanges

Also known as terminal phalanges, these bones play a number of roles. They support the fleshy underside of fingertips in humans, which contain numerous nerve endings. Small, flat extensions from these bones called apical tufts are present underneath nails. A number of adaptations in the anatomy of the distal phalanx of the thumb are necessary for fine motor movement and for grasping objects. In large ungulates such as the giraffe, elongated, large distal phalanges are important in absorbing the stress from running. Similarly, arboreal mammals also have modified distal phalanges.

Distal phalanx bones of the toes taper towards the outside, being broadest near the base next to the proximal phalanx.

Functions of Phalanges

These bones form the structure of fingers and toes and are crucial for most of the evolutionary activity in hominid species. Tool making, precise gripping, grasping and handling of equipment arose due to anatomical features of the phalanges. The opposable thumb appeared due to the evolution of the tendon and ligament structure surrounding the bones. A number of adaptations in the anatomy of the distal phalanx of the thumb are necessary for fine motor movement.

In other animals, phalanges are adapted for quick movement (horses and giraffes), for moving across a canopy (arboreal mammals), for flight (wings in bats, birds), for swimming (fins in aquatic species), and for hunting (claws and paws in carnivores).

Vestigial Phalanges

The evolution of mammals has given the phalangeal formula of 14 bones per limb. In many taxa, however, one or more fingers have lost their immediate function. Dogs, cats, cattle and many other hoofed animals have a reduced thumb that is called a dewclaw. This phalanx is not normally used during movement.

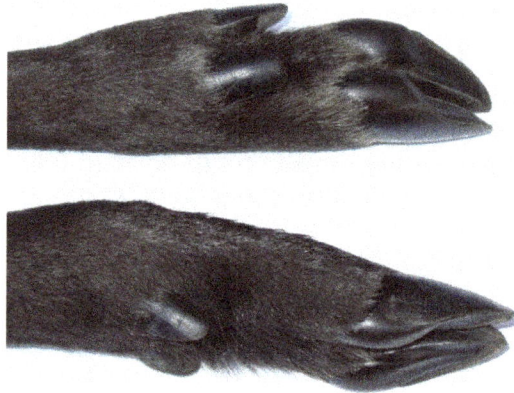

Cloven hoof of roe deer

Occasionally, dewclaws in animals like pigs come into contact with the ground. Reduction of the thumb is common among rats, mice, dormice, kangaroo rats, and squirrels.

In some monkeys, such as the *Colobus* the thumb is vestigial. Many anteaters, armadillos and sloths have vestigial or missing phalanges. The giant armadillo and nine-banded armadillo have vestigial fifth fingers. In the arboreal three-toed sloth, two digits with their accompanying phalanges are not present since they are not important for the animal's lifestyle. The giant anteater has a fifth finger with only 2 phalanges. In lesser anteaters and silky anteaters, the fifth finger has lost all its phalanges.

References

- Clark, Micheal A.; Lucett, Scott C.; Corn, Rodney J., eds. (2008). "Ball Squat, Curl to Press". NASM Essentials of Personal Fitness Training. p. 286. ISBN 978-0-7817-8291-3

- Nguyen, A.K.D (2012). "Head Circumference in Canadian Male Adults: Development of a Normalized Chart". International Journal of Morphology. 30 (4): 1474–1480. doi:10.4067/s0717-95022012000400033

- Bramble, Dennis; Lieberman, Daniel (23 September 2004). "Endurance running and the evolution of Homo". Nature. 432: 345–352. doi:10.1038/nature03052. PMID 15549097. Retrieved 14 November 2014

- Starrett, Kelly; Cordoza, Glen (2013). Becoming a Supple Leopard: The Ultimate Guide to Resolving Pain, Preventing Injury, and Optimizing Athletic Performance. Las Vegas: Victory Belt. p. 391. ISBN 978-1-936608-58-4

- Mester, J. L. (2011). "Analysis of prevalence and degree of macrocephaly in patients with germline PTEN mutations and of brain weight in Pten knock-in murine model". European Journal of Human Genetics. 19 (7): 763. doi:10.1038/ejhg.2011.20

- Farquharson, Claire; Greig, Matt (2015). "Temporal efficacy of kinesiology tape vs. Traditional stretching methods on hamstring extensibility". International Journal of Sports Physical Therapy. 10(1): 45–51. PMC 4325287. PMID 25709862

Permissions

All chapters in this book are published with permission under the Creative Commons Attribution Share Alike License or equivalent. Every chapter published in this book has been scrutinized by our experts. Their significance has been extensively debated. The topics covered herein carry significant information for a comprehensive understanding. They may even be implemented as practical applications or may be referred to as a beginning point for further studies.

We would like to thank the editorial team for lending their expertise to make the book truly unique. They have played a crucial role in the development of this book. Without their invaluable contributions this book wouldn't have been possible. They have made vital efforts to compile up to date information on the varied aspects of this subject to make this book a valuable addition to the collection of many professionals and students.

This book was conceptualized with the vision of imparting up-to-date and integrated information in this field. To ensure the same, a matchless editorial board was set up. Every individual on the board went through rigorous rounds of assessment to prove their worth. After which they invested a large part of their time researching and compiling the most relevant data for our readers.

The editorial board has been involved in producing this book since its inception. They have spent rigorous hours researching and exploring the diverse topics which have resulted in the successful publishing of this book. They have passed on their knowledge of decades through this book. To expedite this challenging task, the publisher supported the team at every step. A small team of assistant editors was also appointed to further simplify the editing procedure and attain best results for the readers.

Apart from the editorial board, the designing team has also invested a significant amount of their time in understanding the subject and creating the most relevant covers. They scrutinized every image to scout for the most suitable representation of the subject and create an appropriate cover for the book.

The publishing team has been an ardent support to the editorial, designing and production team. Their endless efforts to recruit the best for this project, has resulted in the accomplishment of this book. They are a veteran in the field of academics and their pool of knowledge is as vast as their experience in printing. Their expertise and guidance has proved useful at every step. Their uncompromising quality standards have made this book an exceptional effort. Their encouragement from time to time has been an inspiration for everyone.

The publisher and the editorial board hope that this book will prove to be a valuable piece of knowledge for students, practitioners and scholars across the globe.

Index